UNDE STANDING BUSINESS

STATISTICS
MADE EASY

Rob Dransfield

Published in 2003 by:
Nelson Thornes Ltd
Delta Place
27 Bath Road
Cheltenham
GL53 7TH
United Kingdom

03 04 05 06 07 / 10 9 8 7 6 5 4 3 2 1

A catalogue record for this book is available from The British Library.

ISBN 0 7487 7080 1

Diagrams by Steve Ballinger, cartoons by Nathan Betts
Typesetting and page make-up by Paul Manning

Printed in Great Britain by Scotprint

The author would like to thank Jane Cotter, Helen Kerindi and Paul Manning for their help in the preparation of this book.

Thanks are due to the following for permission to use copyright material:
Museum of London (Bills of Mortality, page 2); Press Association (photo of Robbie Williams, page 24).

Contents

Introduction V

How to Use This Book vi

Unit 1: Organising, Presenting and Interpreting Statistical Data

1.1:   Defining Statistics 2
1.2:   Organising, Presenting and Interpreting Information 4
1.3:   Tables and Frequency Distributions 5
1.4:   Pictograms 8
1.5:   Bar Charts 9
1.6:   Pie Charts 10
1.7:   Line Graphs 11
1.8:   Scatter Graphs 13
1.9:   Gantt Charts 15
1.10: Histograms 17
1.11: Frequency Polygons 19
        Questions 20

Unit 2: Averages and Dispersion

2.1:   The Mean 24
2.2:   The Median 26
2.3:   The Mode 27
2.4:   The Range 28
2.5:   Quartiles and the Interquartile Range 30
2.6:   The Standard Deviation 32
        Questions 34

Unit 3: Probability

3.1:   The Probability of Events Occurring 38
3.2:   The Probability of Combinations of Events Occurring 41
3.3:   Expected Values 43
3.4:   Decision Trees 44
3.5:   The Normal Distribution 49
3.6:   Sampling 53
        Questions 55

Unit 4: Relationships Between Variables

4.1:   Scatter Graphs 58
4.2:   Correlation 60
4.3:   Regression 63
4.4:   Forecasting 64
        Questions 66

Unit 5: Time Series

5.1:   Underlying Trends 70
5.2:   Seasonal Variations 74
5.3:   Seasonally Adjusted Figures 78
5.4:   Forecasting Future Figures 80
5.5:   Cyclical and Random Variation 81
       Questions 82

Unit 6: Indices

6.1:   Simple Index Numbers 84
6.2:   Complex Index Numbers 87
6.3:   Important Published Indices 89
       Questions 91

       Answers 93
       Glossary 101
       Index 103

Introduction

This book has been written to give you a clear guide to typical statistical approaches used in business.

Each unit gives you an outline of a major approach or technique used in the gathering and interpretation of statistics. The questions at the end of the unit provide you with the opportunity to test your understanding of what you have read.

Business and economic statistics appear in a variety of forms, but you are most likely to come across them in newspaper and journal articles, in television and radio presentations and in company reports and other business literature. Very often you will find yourself bombarded with statistics. However, reading this book will give you a better appreciation of what to look for in statistical information and will help you to discriminate between useful and not-so-useful information. You will also develop the skills of using statistical approaches to make sense of and interpret raw data.

The book begins by introducing the typical formats in which statistical information is presented. It goes on to examine important techniques for organising and interpreting information.

Doctor Proctor is the helpful guide to statistical approaches. He breaks down the information into bite-sized chunks and uses simple explanations to help you grasp the important components of your statistics course.

At the end of the book you will find a glossary of the more complex terms used and answers to the end-of-unit questions.

About the author

Rob Dransfield is a senior lecturer in business and economics at The Nottingham Trent University. He teaches quantitative methods to undergraduate students, providing a toolkit of statistical approaches which can be applied across a wide range of business and economics courses.

How to Use this Book

As you work through the text, you'll find the following features to help you.

Key Ideas

These are some of the fundamental ideas on which statistics is based.

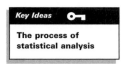

Key Ideas 🔑

The process of statistical analysis

You Must Know This

Terms and principles that you need to learn by heart and understand

Dr Proctor says:
'You Must Know This!'

On a line graph it is normal to show time along the **horizontal** axis and the variables (i.e. items whose value changes) on the **vertical** axis.

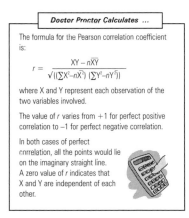

Doctor Proctor Calculates

Learn these methods of calculation – you'll save yourself a lot of time!

Doctor Proctor Calculates ...

The formula for the Pearson correlation coefficient is:

$$r = \frac{XY - n\overline{X}\overline{Y}}{\sqrt{((\sum X^2 - n\overline{X}^2)(\sum Y^2 - n\overline{Y}^2))}}$$

where X and Y represent each observation of the two variables involved.

The value of r varies from $+1$ for perfect positive correlation to -1 for perfect negative correlation.

In both cases of perfect correlation, all the points would lie on the imaginary straight line. A zero value of r indicates that X and Y are independent of each other.

Doctor Proctor Outlines

Explanations of important themes and ideas in statistics.

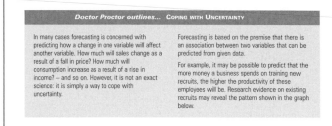

Doctor Proctor outlines... COPING WITH UNCERTAINTY

In many cases forecasting is concerned with predicting how a change in one variable will affect another variable. How much will sales change as a result of a fall in price? How much will consumption increase as a result of a rise in income? – and so on. However, it is not an exact science: it is simply a way to cope with uncertainty.

Forecasting is based on the premise that there is an association between two variables that can be predicted from given data.

For example, it may be possible to predict that the more money a business spends on training new recruits, the higher the productivity of these employees will be. Research evidence on existing recruits may reveal the pattern shown in the graph below.

Distinguish Between...

Here you need to be able to explain the difference between one term or concept and another.

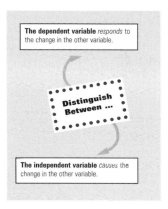

The dependent variable *responds* to the change in the other variable.

Distinguish Between ...

The independent variable *causes* the change in the other variable.

Questions and Answers

Short, practical exercises to test your understanding

QUESTIONS

4.4 Forecasting

1 Give examples of **three** relationships for which a business might want to provide forecasts.
2 How useful would **one** of these forecasts be in giving the business greater levels of control?
3 Why might forecasts not be 100% accurate?
4 If the turnover of a retailer is always twice as much as the cost of sales, what equation would you use to show the slope of the line of best fit?

What is the probability that the next item sold will bring in twice as much revenue as the cost of sale?

5 The chart below shows global market share in $ billions between 1982 and 2002. Is it possible to use this evidence to make forecasts of future music sales during the period 2003–2010?

UNIT 1
ORGANISING, PRESENTING AND INTERPRETING STATISTICAL DATA

Businesses today produce a mass of statistical information – about sales and marketing, production, customers, staff and many other areas of business activity.

In order to be usable, this information needs to be organised, analysed and interpreted. It also needs to be presented in an appropriate way for its intended audience. This unit looks at typical ways of presenting statistical information and analysis.

Topics covered in this unit

1.1 Defining statistics
What is meant by statistics? History and development of statistical techniques.

1.2 Organising, presenting and interpreting information
The process of collating raw data so that it can be logically presented and analysed.

1.3 Tables (including frequency distributions)
Setting out and using different types of table.

1.4 Pictograms
Using a visual approach to make information easier to understand.

1.5 Bar charts
Using bars to represent and compare values.

1.6 Pie charts
A quick and easy way to show how a total can be broken down into different-sized portions.

1.7 Line graphs
Using a line to represent the rise and fall of a variable over time.

1.8 Scatter diagrams
Showing the relationship between two variables in a visual form.

1.9 Gantt charts
How to use a Gantt chart to plan projects and monitor performance.

1.10 Histograms
A more sophisticated form of bar chart in which the area of a bar indicates its relative importance.

1.11 Frequency polygons
How to join together the mid-points of histogram bars to show the frequency of an event or events.

1.1 DEFINING STATISTICS

Key Ideas 🔑

Defining statistics

The term **statistics** refers to the systematic collection and interpretation of data.

Statistics are useful in many ways, but in the wrong hands they can be dangerous. The Victorian Prime Minister Benjamin Disraeli famously remarked, 'There are lies, damned lies, and statistics'. What he meant was that, in quoting statistics, people often use figures to distort, rather than to reveal, the truth.

Politicians are notorious for quoting statistics out of context and for twisting statistical data to support their arguments.

The Italian dictator Benito Mussolini was fond of saying *'Mussolini ha sempre ragione'* ('Mussolini is always right'). He insisted that the government statistics department should always provide statistics that supported his point of view.

Doctor Proctor outlines... EARLY USES OF STATISTICS

One of the most important early statistical surveys carried out in this country was the **Domesday Book**.

Commissioned by William the Conqueror in 1085, the book listed every parish and property in England and was intended to give the Norman king as much information as possible about the country that he had conquered.

However, the true 'founding father' of statistical research was **John Graunt** (1620–74).

Graunt was the author of a work published in 1662 entitled 'Natural and Political Observations Upon the Bills of Mortality'. The 'Bills of Mortality' were the collections of mortality figures in London and consisted of weekly records of deaths due to different causes – for example, 'Murthered' (2), 'Aged' (32), 'Wormes' (11), 'Winde' (3).

By analysing the figures for deaths from plague in Elizabethan times, Graunt was able to provide an early warning system of the threat of fresh plague outbreaks. He also calculated the infant mortality

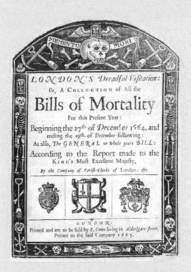

Weekly mortality figures were collected and published in London as early as the seventeenth century.

rate – in this case, the number of children who died before the age of six – and was the first to use representative samples to estimate population growth.

1.1 DEFINING STATISTICS

At its simplest level, statistics involves the gathering and collating of quantitiative information. This can be termed **descriptive statistics**.

But an important task of the statistician is to make what are known as **statistical inferences** – i.e. to make deductions based on the information that has been collected.

Descriptive statistics are used to summarise or describe observations. For example, we may say that 'sales increased by 50% this year', or that 'young people visit the cinema more frequently than old people'.

Distinguish Between ...

Which of these headlines would you say is **descriptive** and which is **inferential**?

Record increase in small business bankruptcies last year

Experts predict autumn slowdown in economy

Inferential statistics use observations as a basis for making estimates or predictions about the future. For example, we may say that 'if share prices continue to rise, they will reach record highs by September next year'.

Example

One of the earliest recorded uses of **statistical inference** was by John Graunt in his studies of infant mortality in the seventeenth century.

From the various causes of death listed in the Bills of Mortality *(see opposite)*, Graunt isolated those that applied specifically to children. To these he added half the deaths from smallpox and measles, which, using his own observations, he assumed to be of children. He added a further one-third of all plague deaths to this total.

Using the figures for infant baptisms, and the figures that he had deduced for child deaths, Graunt was thus able to make a statistical inference of infant mortality.

1.2 ORGANISING, PRESENTING AND INTERPRETING INFORMATION

Key Ideas 🔑

The process of statistical analysis

Before information can be analysed, it needs to be collected and sorted.

For example, information about consumers' views on a product needs to be collected by market research *(below)*.

Statistics begins with the collection of **raw data**. This is obtained by collecting observations or measurements. Data can consist of non-numerical qualities such as shapes and colours of different products, or it may be made up of numerical quantities such as costs and sales figures.

All data must be stored in a databank or catalogued as a list. An **inventory** is a document setting out a basic set of data from which different types of **inferences** can be made.

In most situations the first step of analysis is to determine the **frequency distribution** of the collected data. This can refer to either:

- the frequency of occurrence of each listed value or quality, *or*
- the frequency of occurrence of values or numbers which are in a certain **interval** (also termed a **class** or **fraction**).

Doctor Proctor outlines... RESEARCHING THE MARKET

Market research provides researchers with raw data. This data needs to be sorted and analysed.

For example, in a survey carried out by a bakery to find out which of two cakes customers preferred, participants were asked the following question:

> *Which of the two cakes that you have just tried do you prefer ? (tick the box)*
>
> *Sample A* ☐
> *Sample B* ☐
> *No preference* ☐

When the results were collected, the survey showed that of 50 people surveyed, 40 preferred sample A, 8 preferred sample B, and 2 had no preference.

1.3 TABLES AND FREQUENCY DISTRIBUTIONS

Key Ideas

Creating a linear array

Once data has been collected, there are three important steps in making sense of statistical information. These are:

- **interpreting**: establishing what the information means
- **presenting**: choosing the best way of showing the information
- **organising**: assembling the information in a structured way.

If the number of observations collected is small (e.g. the sales figures for a new book in the first ten days after publication), the data can be set out in the following way:

50　60　74　120　180　62　75　51　123　96

The same data could be arranged in ascending or descending magnitude:

50　51　60　62　74　75　96　120　123　180

Alternatively, as there is only one observation for each variable, the information could be displayed on a graph in the form of a linear array of dots:

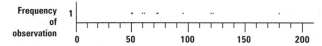

If there is a second observation of any of the variables – e.g. if there were two days on which sales were 120 – the dot at 120 would appear at twice the height.

Because there are so few observations in this example, there is very little statistical value to it. However, it is possible to identify the **range** of the observations, i.e. the difference between the highest and lowest values. This is worked out by subtracting 50 (the lowest number) from 180 (the highest number). The range is therefore **130**.

Doctor Proctor outlines... COLUMNAR LAYOUT

A **table** is a set of facts or figures that are set out in an ordered way, typically in columns.

For example, the table *(right)* compares the finances of Manchester United and Charlton Athletic for 2002.

The table allows us to make some interesting comparisons. For example,

1 Manchester United has a much higher income than Charlton.

2 Manchester United's income from playing in the Champions' League is greater than Charlton's entire annual income.

3 With its higher income, Manchester United is able to spend more money on expensive 'star' players.

Financial Comparison		
	Charlton Athletic	Manchester United
	£m	£m
TV income	13.88	25.82
Champions' League income	–	34.80
Season ticket income	7.20	28.00
Annual shirt sponsorship	1.10	7.50
Annual merchandise income	1.27	21.90
Record transfer fee paid	4.75	30.00
Wages	17.07	50.00

1.3 TABLES AND FREQUENCY DISTRIBUTIONS

Key Ideas

Frequency tables

So far, the examples we have looked at have been relatively simple. But as the number of observations increases, we find we need to arrange the data in a more manageable form.

For example, in an office where employees work flexitime, 100 members of staff work the following daily hours:

5	8	7	9	2	5	6	8	9	6
7	5	8	6	4	5	9	9	9	9
6	4	8	8	9	9	9	8	7	2
5	8	8	9	9	7	9	8	8	6
3	7	7	7	4	9	4	7	4	8
8	9	9	9	7	6	4	9	9	9
8	9	9	8	8	4	6	5	8	8
9	8	9	2	5	5	5	7	8	7
8	9	9	7	7	5	5	6	6	6
9	8	8	8	7	7	9	7	9	9

In this form, the data is of little use. It would be much more useful to display it in what is known as a **frequency table**.

In the table below, the top line shows all the values of the variable (the number of hours worked per employee). On the bottom line is the number of times that each value of the variable was observed (i.e. the frequency):

Frequency Table showing Number of Hours Worked by Employees

Number of hours worked	2	3	4	5	6	7	8	9
Frequency of observation	3	1	7	11	10	16	23	29

Reading from the table, it is easy to see that 3 employees worked 2 hours, 1 employee worked 3 hours, and so on.

You can also calculate the total number of employees who were observed. This is done by adding together all the frequencies:

$$3 + 1 + 7 + 11 + 10 + 16 + 23 + 29 = 100$$

Doctor Proctor outlines... RELATIVE FREQUENCY

In the frequency table above we can see that 10 employees worked for 6 hours. But is that a large number or not? In order to answer that question we need to set out a **relative frequency table**.

The formula for relative frequency is:

$$\text{Relative frequency} = \frac{\text{Actual frequency}}{\text{Sum of all the frequencies}}$$

Thus the relative frequency of employees working six hours is:

$$\frac{10}{100} = \quad 10\%$$

The complete set of data for relative frequencies can be set out as shown *(right)*:

Relative Frequency Table showing Number of Hours Worked per Employee

Number of hours worked	Relative frequency	Percentage
2	0.3	3%
3	0.1	1%
4	0.7	7%
5	1.1	11%
6	1.0	10%
7	1.6	16%
8	2.3	23%
9	2.9	29%

The table shows us that the most frequently observed numbers of hours for employees are 8 (23%) and 9 (29%).

1.3 TABLES AND FREQUENCY DISTRIBUTIONS

Key Ideas

Making comparisons

One of the great advantages of the table format is that it allows comparisons to be made. This is a feature of all the types of table we have looked at so far.

Take a look at the table below showing average national salaries over the period 1992–2001.

What are the key comparisons that the table brings to light? Which groups have risen up the pay league table and which groups have fallen during the last ten years?

Example

AVERAGE NATIONAL SALARIES 1992–2001

Rank 2001	Rank 1992	Occupation	Annual Salary 2001 (£)	Annual Salary 1992 (£)	Change (%)
1	(1)	General managers (large companies)	110,341	52,449	110.4
2	(8)	Barristers	78,549	30,813	154.9
3	(3)	Senior civil service administrators	61,993	39,994	55.2
4	(4)	Company financial managers	59,129	25,692	130.1
5	(2)	Aircraft pilots	57,328	42,055	36.3
6	(6)	Managers in mining and energy	48,916	32,017	52.8
7	(7)	Doctors	48,235	30,822	56.5
8	(11)	Insurance underwriters, claims assessors, brokers	46,034	28,928	59.1
9	(10)	Management consultants	45,313	29,914	51.5
10	(12)	Judges and court officers	45,022	28,435	58.3
11	(16)	IT managers	43,268	27,440	57.7
12	(21)	Work study managers	42,867	26,099	64.2
13	(9)	Police officer (inspector and above)	41,984	30,660	36.9
14	(19)	Chemical engineers	40,926	26,657	53.5
15	(5)	Air, commodity and ship brokers	40,421	33,144	22.0
16	(14)	Solicitors	39,775	28,081	41.6
17	(24)	Marketing and sales managers	39,750	25,692	54.7
18	(25)	Personnel managers	37,216	25,187	47.8
19	(23)	Physicists, geologists and meteorologists	36,829	25,890	42.3
20	(15)	Air traffic, planning and controllers	36,797	24,476	33.9

1.4 PICTOGRAMS

Key Ideas

Visual representation

Numbers of Male and Female Directors in Two Contrasting Companies

	Male directors	Female directors
Topdown Oilco	8	1
Democratic Mediaco	5	5

A **pictogram** uses images to make information easier to understand.

For example, look at the table *(above right)* showing the number of male and female directors in two contrasting companies. Then look at the pictogram below it, based on the same information. Which do you think gets the message across more effectively?

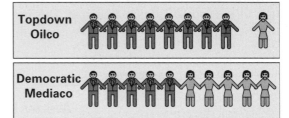

Pictograms are very useful for making comparisons in a visual way. They are often used to summarise information in company reports or multimedia presentations.

For example, this pictogram shows the different sectors of the UK confectionery market in 2002.

Sales of Confectionery in the UK, 2002

Bitesize (e.g. Smarties) 10%
Easter eggs 5%
Boxed chocolates 10%
Blocks (i.e. bars) 60%
Other 15%

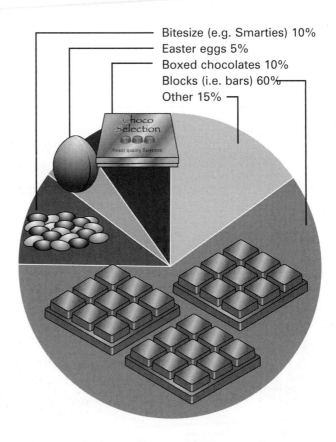

1.5 BAR CHARTS

Key Ideas 🔑

Visualising relative proportions

Often in survey work it is important to see the relative proportions of data in each category. In the table below, the data from a supermarket survey has been converted into percentages and arranged in order of size:

Transport Used by Supermarket Shoppers

Means of transport	% of users	Actual numbers in sample
Private car	50	25
By foot	24	12
Taxi	16	8
Train	4	2
Motorbike	4	2
Cycle	1	1
	Total =	**50**

You can see that the table is much easier to read and interpret than raw data. But to make the picture clearer still, we can set the information out in a **bar chart**, like this:

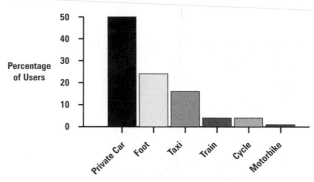

A bar chart is a diagram in which bars have lengths which represent quantities or values. Bar charts are the most popular way of representing statistical information because they are simple to construct and read. Microsoft Powerpoint lets you produce a simple bar chart in moments.

When creating a bar chart it is important to identify the measurements you are using and to use clear labelling and a key where necessary.

Bar charts can be set out in either a vertical or a horizontal format. A vertical bar chart is best when a relatively small amount of text is needed to explain each of the bars. When more detailed explanation is needed, a horizontal format works better.

The horizontal chart below is based on a survey carried out by *The Face* magazine in 2002 into the views of 11,000 people between the ages of 16 and 29.

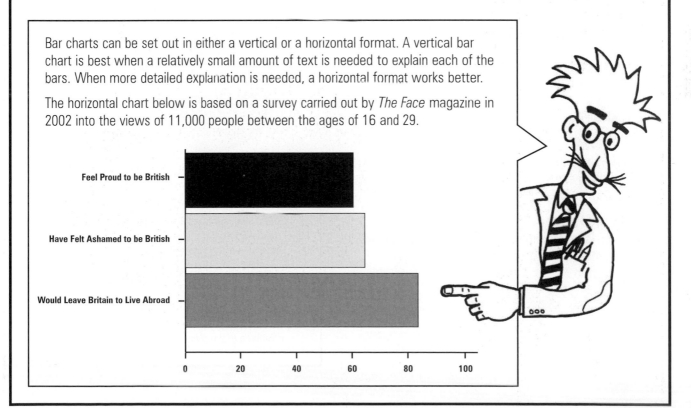

1.6 PIE CHARTS

Key Ideas 🔑

Segments

One of the simplest and most effective ways to illustrate relative frequency is to use a **pie chart**.

In a pie chart, a circle is divided up into different-sized segments, to represent the proportion of the categories to the total. For example, if the value of a firm's sales is shared equally between two products, the circle would be divided exactly in half and each product would be represented by a segment of 180^0.

In a pie chart, the angle that each segment makes at the centre of the circle is proportional to the frequency of the category concerned.

Doctor Proctor outlines... CREATING A PIE CHART

Today most business reports and presentations are produced on computer. Pie charts can be created very quickly and easily with a range of colours and different types of shading. If you enter the values for each of the categories, the computer will calculate the size of the segments and display the results automatically.

For example, the pie chart *(right)* shows the relationship between the turnover (e.g. from gate receipts, sale of club souvenirs, etc.) of Sunderland Football Club in the six months to 31 January 2002, and the profit made from these sales.

The value of turnover was £25 million, and the operating expenses of the club were £20 million, leaving profits of £5 million.

The total shown in the pie chart is £25 million, which is represented by 360^0. Each £1 million will therefore be represented by 14.4^0, which is calculated by:

$$\frac{360^0}{25} = 14.4^0$$

The operating expenses segment will make up 80% of the pie – i.e. 288^0. The remaining 72^0 of the pie represents profit.

Sunderland Football Club
August 2001 – January 2002

Operating Profit

Operating Expenses

Uses of pie charts

*   To show who gets what proportion of a particular 'cake' – for example, how the profits of a company are shared among stakeholders.

*   To show how the total sales of a firm can be broken down by product, production process, location, etc.

*   To show how different costs, or types of costs, contribute to the overall price of a product.

1.7 LINE GRAPHS

Key Ideas 🔑

Values which change over time

Line graphs are particularly useful in showing how values or quantities rise and fall over a period of time.

Graphs showing changes in the value of variables over time are also called **time series** charts. These types of chart are often used to present information in business reports, especially financial statements and company reports.

Dr Proctor says:
'You Must Know This!'

On a line graph it is normal to show time along the **horizontal** axis and the variables (i.e. items whose value changes) on the **vertical** axis.

Uses of time series charts

Time series charts can be used for showing a wide range of economic variables:

- interest rates
- unemployment rates
- inflation rates
- exchange rates
- the rate of growth of the economy
- changes in population
- fluctuations in share prices.

For example, the chart *(right)* shows the share price of the music, record company and publishing group EMI between September 2001 and August 2002. The fall in the share price from the spring of 2002 resulted from general economic pessimism, coupled with gloom about the potential of the music industry following a steady fall in sales of CDs.

Illustrating patterns and trends
Time series charts are often used to illustrate trends. The graph *(right)* shows improvements in the A Level pass rate over time.

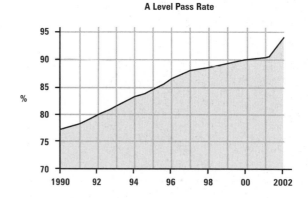

Key Ideas

Comparing sets of variables

The graph *(below)* illustrates changes in the market share of two rival companies and allows the two sets of figures to be compared.

Line graphs can also be used to make comparisons between two or more sets of observations.

Graphs like those shown below are often used in the financial press to illustrate the relationship between economic factors.

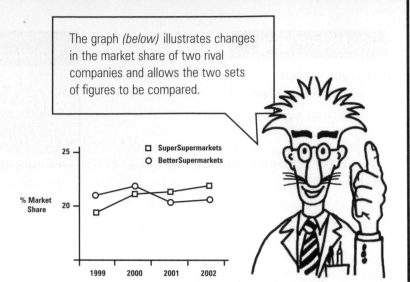

Example

When the economy is in a downturn, more people lose their jobs, leading to an increase in unemployment. Rising unemployment means that people have less money to spend – which in turn leads to a fall in the general level of prices and therefore to a fall in inflation.

The graphs *(left)* illustrate this trend in the period from 1990 to 'Black Wednesday', 16 September 1992, when currency speculators forced Britain to leave the Exchange Rate Mechanism.

The relation can be shown even more clearly when the two sets of time series figures are plotted on the same chart *(below)*.

1.8 SCATTER GRAPHS

Creating a scatter graph simply involves plotting a series of points for X and Y values. Once the points have been plotted, the graph will show whether a relationship can be identified between the two.

The simplest way of studying the relationship between two variables is to use a scatter graph.

The data for a scatter graph is drawn from a single sample of subjects (e.g. individuals or business units), with two measurements being made for each subject. Each individual's measurements are plotted as a single point, with the X measurement plotted on the horizontal axis and the Y measurement plotted on the vertical axis.

Example

In order to examine the success of Premier League Football businesses in the first ten years of the Premier League, we might want to examine the record of the six most successful football clubs. We could start with the raw data set out in the table below.

The First Decade of the Premiership (1992–2002)

		Played	Won	Drawn	Lost	Goals For	Goals Against	Points
1	Manchester United	392	244	93	55	789	360	825
2	Arsenal	392	195	110	87	598	346	695
3	Liverpool	392	189	100	103	643	408	667
4	Leeds	392	167	111	114	543	437	612
5	Chelsea	392	160	118	114	589	451	598
6	Aston Villa	392	154	115	123	491	434	577

We could then set out a scatter graph *(below)* showing the relationship between goals scored and number of matches won.

The scatter graph suggests that there *is* a relationship between the number of goals scored and the number of games won – as represented by the straight line.

But it is also clear that 'number of goals scored' does not tell the whole story. For example, Arsenal have won more matches than Liverpool because the evidence shows that they have a better defensive record.

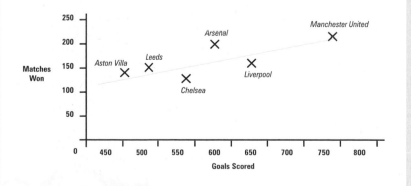

1.8 SCATTER GRAPHS

Example

One of the best known scatter graphs was produced by the Boston Consultancy Group.
The Group examined two variables in relation to business in America:

- the total output produced by firms
- the cost of producing units of output.

The Group noticed that over time, the cost of producing units of output decreased as a result of the 'experience' that large firms acquired. They developed the hypothesis that with each doubling of experience, average costs per unit tend to fall by 20–30%.

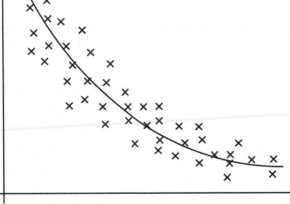

The research carried out by the Boston group showed that profits of a company are related to market share, as follows:

Market Share %	Profitability%
Under 7	9.6
7–14	12.0
14–22	13.5
22–36	17.9
Over 30	30.2

The cost of producing each unit falls as total output increases over time. This makes sense: the more you do something, the more experience you have of doing it, and the more likely you are to do it better because you can identify faults and take steps to correct them. Gains from efficiency stem from greater experience. Large firms can also benefit from a range of economies of scale such as bulk buying.

1.9 GANTT CHARTS

> **Key Ideas** 🔑
>
> **Charts as planning tools**

Gantt charts are a form of bar chart developed by the management planner, **Henry Gantt** (1861–1919). They are especially useful as a visual tool for project planning and performance monitoring.

Gantt charts are designed to show the relationship between different tasks. Using Gantt charts, managers can check actual progress against planned progress and spot when hold-ups in a particular task are likely to slow down the completion of the project as a whole.

Gantt charts can be drawn manually or produced easily using a graphing package.

Doctor Proctor outlines... CONSTRUCTING A GANTT CHART (1)

To draw a Gantt chart for a project, start by breaking it down into distinct tasks or activities and estimating the time needed to carry out each one.

Next, identify the relationships between the tasks.

Here are some special terms used in project planning:

- The earliest **start time** of an activity means the earliest time that it can commence. This will usually depend on the earliest finish of the preceding activity.

- The earliest **finish time** for an activity is the earliest start + the **duration** of the activity.

- **Slack** is the surplus time that is available between the earliest time that an activity can be completed and the latest time that it can be completed if the project is to run on schedule. If there is no slack time available for an activity, it must be given priority and special care must be taken to ensure that it runs to time. If slack time *is* available, it is important to check that it is not wasted – even activities with slack time need to be monitored closely.

Activities that must be carried out in a **serial relationship**, i.e. **C** can only be carried out once **B** is completed, and **B** can only be completed after **A**:

$$A \longrightarrow B \longrightarrow C$$

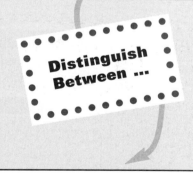

Distinguish Between ...

Activities that can be carried out at the same time because they are independent of one another, i.e. **parallel relationships**:

$$A \longrightarrow$$
$$B \longrightarrow$$
$$C \longrightarrow$$

1.9 GANTT CHARTS

Doctor Proctor outlines... CONSTRUCTING A GANTT CHART (2)

1   Draw a grid. Plot the tasks along the vertical axis, and the timescale (up to the end of the project) along the horizontal axis.

2   Draw a horizontal bar from the description of the task on the left of the chart, starting at the earliest start time and ending at the earliest finish time.

3   Show the slack amount by drawing a line from the earliest finish time to the latest finish time.

4   Repeat steps 2 and 3 for each task.

Take the example of a project consisting of eight main tasks:

• **Task A** must be carried out at the start of the project and is expected to take 30 days.

• **Task B** must also be carried out at the start of the project and is expected to take 15 days.

• **Task C** follows **B**. Duration: 30 days.

• **Task D** follows **A** and **C**. Duration: 15 days.
• **Task E** follows **C**. Duration: 10 days.
• **Task F** follows **E**. Duration: 14 days.
• **Task G** follows both **A** and **C**. Duration: 7 days.
• **Task H** follows **D**, **E**, and **G**. Duration: 14 days.

Once **H** is completed the project is finished.

It is calculated that the project can be carried out in 83 days.

This information can be plotted onto a Gantt chart as illustrated below:

Note that the earliest start and finish times for each activity are drawn in a horizontal bar. The slack times are shown for relevant activities with a line extending to the latest finish time that would be acceptable if the project is to finish on time, i.e. 83 days. (You can see, for example, that **A** doesn't necessarily have to be completed until the time when **D** must start. **F** can only start after **A**, **B** and **C** are completed.)

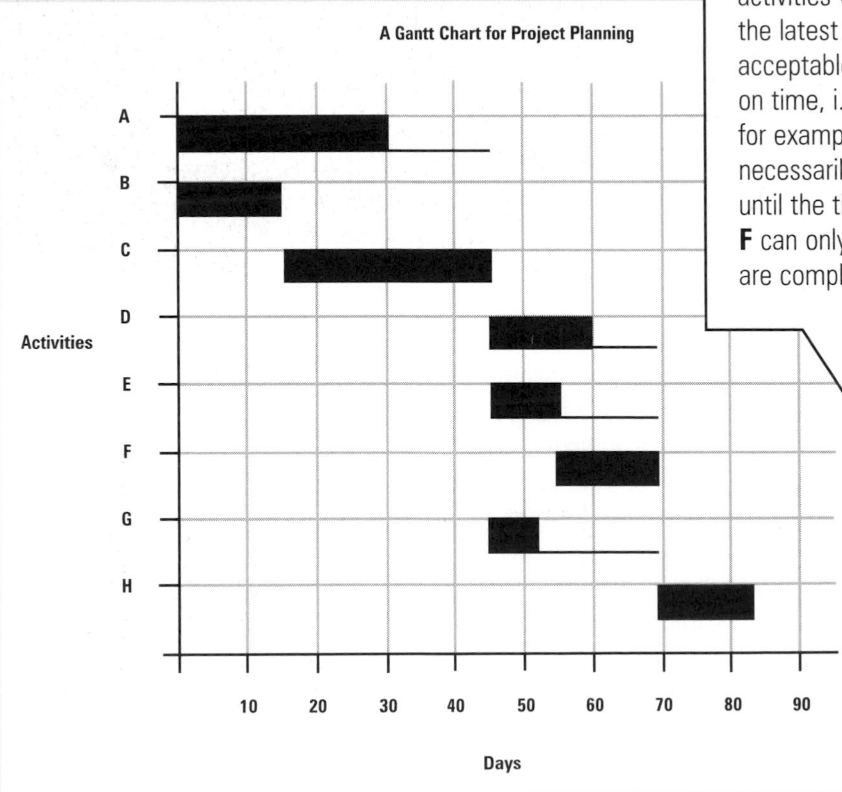

A Gantt Chart for Project Planning

Activities: A, B, C, D, E, F, G, H

Days: 10, 20, 30, 40, 50, 60, 70, 80, 90

1.10 HISTOGRAMS

Key Ideas

Variables and frequency

A histogram is a block diagram whose blocks are proportional in area to the frequency in each class or group.

The example below shows a typical histogram with equal divisions along the horizontal axis, where the value of the variable is represented by the height of the vertical column.

Dr Proctor says:
'You Must Know This!'

In a histogram the variable is always plotted on the **horizontal** axis and the frequency is always plotted on the **vertical** axis. Equal steps are marked off along the horizontal axis to represent the values of the variable, and the number of observations of each value of the variable is represented by the height of a vertical column.

Example

Suppose that we are carrying out some market research to find out how long it takes viewers to recognise the brand name of a well-known cola drink in a television commercial.

In this case the **variable** is the number of seconds taken to recognise the brand name. The **frequency** is the number of observations at a particular second.

The table *(below)* represents this data:

Seconds Taken to Recognise Cola Drink

Number of observations	Number of seconds
10	1
10	2
20	3
20	4
30	5
10	6

This information is represented in the histogram *(right)*.

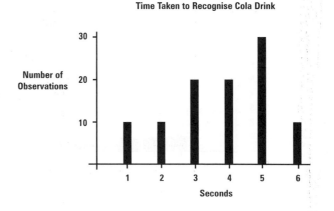

Time Taken to Recognise Cola Drink

1:10 HISTOGRAMS

Key Ideas

Histograms with unequal class intervals

Most histograms are set out in the way outlined on page 17, with each of the bars having an equal width.

However, there may be times when we want to show observations involving different changes in the size of the variable. An example of a histogram based on unequal class intervals is shown below.

Example

In collecting information about the ages of workers in a manufacturing plant, a company has collected details of employees in the age ranges 18–24, 25–34, 35–44, 45–60, 61, 62, 63, 64 and 65. The data have been broken down in this way because the company needs to know when it will have to replace workers who are retiring. Thus the data cover a range of different-sized class intervals: 10 years for 24–34, and 35–44, 1 year for 61, 62, 63, 64 and 65.

The table below shows the frequency for each of these class intervals:

Age Distribution of Workforce

Class	Number
18–24	8
25–34	10
35–44	10
45–60	10
61	3
62	4
63	2
64	4
65	1

If this information were presented on a normal bar chart, it would appear that there were a disproportionate number of employees in the age groups 18–24, 25–34, 35–44 and 45–60. This is inaccurate, because within the group 61–65 there are more employees – 14 – than in any of the other categories.

In setting out a histogram we can give a more accurate picture by averaging out the number of workers in those blocks covering larger class intervals. The area of the block is what we need to look at, rather than just the height of the block.

The histogram reveals that there is a problem in relation to the age of the workforce, because the company has a disproportionate number of workers in the 60–65 age range.

In a histogram, you need to consider the *area* of a block rather than just its height.

1.11 FREQUENCY POLYGONS

Connecting interval midpoints

A **frequency polygon** is often used to show frequencies rather like a histogram.

However, instead of placing a bar over the class interval, it is simply constructed by drawing a dot at the midpoint of the interval. The dots are then joined together. Sometimes the polygon is left open at the ends. Normally it is closed by drawing a straight line from each end dot down to the horizontal axis. The points on the horizontal axis that are chosen to close the frequency polygon are the midpoints of the first class interval (at each end of the distribution) that have a zero frequency.

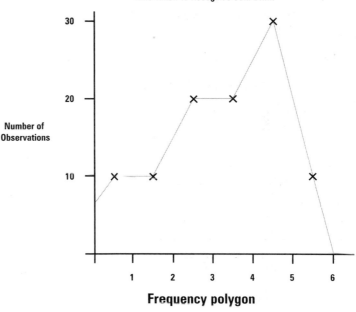

Time Taken to Recognise Cola Drink

Frequency polygon

The frequency polygon *(above right)* simply shows the number of seconds it takes a sample of people to recognise the brand name of a cola drink in a TV advertisement.

The same information could also be set out in the form of a **cumulative frequency polygon**, based on the table below:

Cumulative frequency of observation	Number of seconds
10	1
20	2
40	3
60	4
90	5
100	6

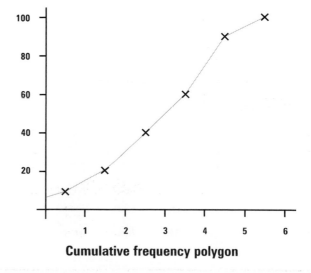

Cumulative frequency polygon

1.2 Organising, presenting, interpreting

Classify the following statements as **descriptive** or **inferential** in their use of statistics.

1 Measurements indicate that the hole in the ozone layer is starting to close.
2 Using these measurement it is possible to estimate how fast the ozone layer will close each year in the future.
3 If stock market prices continue to fall, it is likely that this could trigger a crash in the future, and it is possible to estimate the date when this will happen.
4 Sales have increased by 5% a year for the last five years.

1.3 Tables

The data below shows the average amount spent in a supermarket by 80 shoppers questioned in a market research survey. Organise the figures into a frequency table, using the following classes: £0–£20, £21–£50, £51–£100, £101–£200, over £201.

15	20	5	150	248	12	120	310
8	154	204	12	302	58	112	56
203	16	28	75	150	210	15	32
6	8	15	89	102	250	210	85
76	82	53	45	281	18	38	112
67	48	138	7	100	120	353	46
212	28	38	73	103	103	45	12
83	38	290	238	10	23	65	72
76	76	38	38	36	128	38	200
78	74	46	12	4	283	12	38

1.4 Pictograms

What conclusions can you draw from this pictogram showing sales of wine in Britain by the French exporter Vin Extraordinaire?

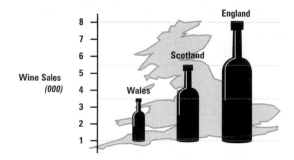

1.5 Bar charts

The bar chart below shows average years of schooling in a number of countries.

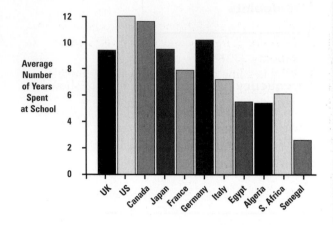

1 How useful is the bar chart in making comparisons between the countries shown?
2 What are the most striking comparisons?
3 Why is a bar chart a good way of displaying this information?

1.6 Pie charts

1 In a survey of 1,000 young people aged between 16 and 29 years, 250 stated that the drug Ecstasy should be decriminalised, while 75% disagreed with this view. Show how this information could be presented in a pie chart.
2 Use a graphing package to show the following information in a pie chart.

Platform Penetration of UK Households – Forecast for 2005

Analogue TV	35%
Digital satellite	30%
Digital terrestrial (non-pay)	18%
Digital cable	17%

3 Why might a pie chart be a better way of displaying this information than a bar chart?

Continued on page 21

1.7 Line graphs

The graph below shows the price of shares in the furniture retailer MFI between January 2001 and March 2002.

1 When was the high point for shares in MFI?
2 How much has the share price risen since January 2001?
3 Why is a line graph a good way of showing changes in share prices over time?

MFI: Share Price Performance Since January 2001

1.8 Scatter graphs

Draw a scatter graph to show the relationship between games won and the defensive record of Britain's top six football clubs during the first ten years of the Premiership. Does there seem to be any relationship?

The First Decade of the Premiership (1992–2002)

	Matches Won	Goals Against
Manchester United	244	360
Arsenal	195	346
Liverpool	189	408
Leeds	187	437
Chelsea	160	451
Aston Villa	154	434

1.9 Gantt charts

A company has identified the following activities that will make up a project:

Activity	Duration (Days)	Immediate Predecessor(s)
A	1	–
B	1	A
C	3	B
D	2	B
E	2	D
F	1	E
G	3	C
H	2	E, F, G
I	1	H
J	1	I
K	1	I
L	1	J, K

Show this information on a Gantt chart. Explain why a Gantt chart is a useful way of displaying this information.

1.10 Histograms and 1.11 Frequency polygons

The table (*below*) shows the number of different items stored in the aisles of a large supermarket, and the frequency of observation of these numbers.

For example, 10 aisles were observed, each displaying 19 items, including those for confectionery items and cleaning products.

Number of items per aisle												
19	20	21	22	23	24	25	26	27	28	29	30	31
Frequency of observation												
10	12	5	14	8	7	12	14	15	18	13	20	12

1 Set out the above information as a histogram.
2 Set out the above information as a frequency polygon.
3 Which do you feel is the best way of displaying the information? Why?

There are several ways of looking at averages. In this unit we look at the difference between the three principal approaches to averaging: the mean, median and mode.

We then go on to examine ways in which values in a data set can be spread (or **dispersed**) around the average.

Topics covered in this unit

2.1 The mean
The most commonly used way of calculating an average.

2.2 The median
In a set of statistics arranged in order of magnitude, the **median** is the value that occurs in the middle of the list.

2.3 The mode
The most frequently occurring value in a set of statistics.

2.4 The range
The difference between the highest and lowest value in a set of statistics. For example, over a year, a share price may fluctuate between 500 and 550 pence (a range of 50 pence), while another may be more volatile, fluctuating between 250 and 750 pence (a range of 500 pence).

2.5 Quartiles and interquartile range
A **quartile** represents a quarter or 25% of a range. The **interquartile range** is the difference between observations for the first quartile and the final quartile.

2.6 The standard deviation
When dealing with quantity variables, we need to be able to measure the extent to which the observed values are spread out from the centre. The most commonly used measure of dispersion is known as the **standard deviation**. This is the 'average' amount by which all the values deviate from the mean.

2.1 THE MEAN

Key Ideas 🔑

Distorting 'averages'

Politicians often use figures to win arguments and score points.

In late 1999, the deputy prime minister John Prescott argued that 'average' full-time earnings in the UK were £20,861 a year. Although this sounded impressive, at the time two-thirds of people in full-time jobs actually earned *less* than the average.

The so-called 'average' figure was distorted by the fact that a relatively small number of people in the UK earn much more than the rest – for example, Premiership footballers, Robbie Williams and cabinet ministers (who at the time earned about £100,000).

High earners like Robbie Williams can distort figures for 'average income' which are based on the arithmetical mean

In this case, the **mean,** or arithmetical average, is not a particularly useful or reliable indicator of 'average' earnings. A better approach would be to use the **median** – the middle figure between highest and lowest earnings. This would have brought the average figure down to £17,360.

Doctor Proctor outlines... CALCULATING THE ARITHMETICAL MEAN

When explaining the difference between the mean, median and mode, it is helpful to use the same set of figures. Here is an array of figures showing the number of cars owned by 10 cabinet ministers.

2   6   4   3   5   6   3   4   3   2

In total, the 10 cabinet ministers own 38 cars.

The most common way of deriving an average, the **mean**, involves simply adding up all the values in a series and then dividing the total by the number of observations.

$$\text{The arithmetic mean} = \frac{\text{Sum of observations}}{\text{Number of observations}}$$

In the case of cabinet ministers' cars, this works out as:

$$\frac{2 + 6 + 4 + 3 + 5 + 6 + 3 + 4 + 3 + 2}{10}$$

= 3.8 cars

Uses of the mean

- Unlike the **median** and the **mode** *(see pages 26 and 27)*, the mean takes into account all items in the data set. Where there are no extreme values, it is a very useful method of averaging. It is widely understood and can be quickly calculated on a computer or calculator.

- The disadvantage of the mean is that it is distorted by extreme values in the data set. It does not lend itself to graphical display, and it may not correspond to any of the actual values in a data set. For example, none of the cabinet ministers above actually owns 3.8 cars!

2.1 THE MEAN

Key Ideas 🔑

Uses of the mean

Typical uses of the mean as an average in business and economics include calculations of:

- average hourly output per unit of input (e.g. machinery or labour)
- average sales per period of time
- average sales per salesperson
- average number of customer complaints per day
- average price increases (as measured by, for example, the Retail Price Index)
- average changes in stock market prices.

The example below explains the process of working out a business's average output over a 50-week period.

In the past, this task was time-consuming, tedious and prone to error.

Nowadays, figures are continuously updated on computer and averages can be calculated automatically.

Example

Over a 50-week period a business's output figures are as follows:

Output (000s)

5	6	4	6	2	5	6	2	6	5
5	6	5	6	6	5	3	5	4	5
5	3	2	3	3	4	2	3	6	5
4	1	5	2	4	1	2	5	3	4
5	4	2	5	3	5	4	4	4	5

These figures can be set out in a frequency distribution table as shown below. The Greek letter sigma ($\sum$) means 'sum of'.

Weekly output level (000)	Frequency of occurrence	Output level x Frequency
x	**f**	**fx**
1	2	2
2	7	14
3	7	21
4	10	40
5	16	80
6	8	48
	$\sum f = 50$	$\sum fx = 205$

Calculating the mean

To work out the arithmetic mean – in this case, average weekly output – simply multiply the weekly output figures by the frequency with which they occur, and then divide by the total number of weeks (50).

$$\frac{205,000 \text{ units}}{50 \text{ weeks}} = 4,100 \text{ units per day}$$

The formula for the arithmetic mean using a frequency distribution table is therefore:

$$\text{Arithmetic mean} = \frac{\sum fx}{\sum f}$$

2.2 THE MEDIAN

The **median** is the middle value in a data set.

For example, take our data set for cabinet ministers and their cars:

2   6   4   3   5   6   3   4   3   2

In order to calculate the median, we must first rearrange these data into order of value:

2   2   3   3   3   4   4   5   6   6

As this is a data set with an even number of items (10 cabinet ministers), there are two middle values, so it is necessary to take a simple average of the two values:

2   2   3   3 | 3 | 4 | 4   5   6   6

$$\text{Average} = \frac{3+4}{2} = 3.5 \text{ cars per cabinet minister}$$

Doctor Proctor outlines... CALCULATING THE MEDIAN

Where many numbers are involved, we can use a frequency distribution table to calculate the median.

Let us go back to the example of weekly output figures. These were set out in a frequency distribution table as follows:

Weekly output (000s)	Frequency of occurrence (f)
1	2
2	7
3	7
4	10
5	16
6	8
	$\Sigma f = 50$

We use the letter n for the total number of frequencies. In this case $n = \Sigma f = 50$.

The middle (median) value is thus $\frac{50}{2} = 25$

– in other words the 25th number in the series, which in this case is 4,000 units of output.

When calculating the median with an uneven total frequency, for example 51 weeks rather than 50, we use the formula:

$$\text{Median} = \frac{(n+1)}{2}$$

Uses of the median

The median is useful for measuring:

- average incomes and earnings
- average hours worked by employees
- average sales made by employees.

It is a useful way of averaging because it avoids the problem of distortion associated with the arithmetic mean. The median is also an actual value which is readily obtained, even if not all the values of the items are known.

A disadvantage is that, although the median gives the value of only one item (the middle), a number of the surrounding items may have the same value. If these items are spread erratically above or below it, the median may lose its value as a representative central figure.

2.3 The Mode

The **mode** is the value that appears most frequently in a distribution. Because it is the most 'popular' value, it can be seen as another way of indicating a typical or representative value among those recorded.

Doctor Proctor outlines... Calculating the Mode

Once again, take the number of cars held by cabinet ministers:

2   6   4   3   5   6   3   4   3   2

You can immediately see that 3 is the most popular value, appearing three times in the data set.

The mode is often most useful when the sample being studied concerns categories rather than quantity variables.

Uses of the mode

The mode can be very useful in forecasting. For example, by identifying the most popular value for monthly sales, it is possible to forecast likely future sales. However, using the mode as a predictor is more likely to be effective in static situations, i.e. those where sales do not change much from month to month, rather than in dynamic situations where there is fluctuation.

Further advantages of using the mode are that it represents an actual recorded value and is not distorted by extreme values. Modes are also easy to understand and can be shown graphically.

In multimodal situations with a widespread distribution, the mode loses its value as an average. For example, take the case of a survey in which 100 viewers are asked to rate a new film from 10 (Excellent) to 1 (Very Poor). If the results are bi-modal, with 30 rating the film 9 and 30 rating it 2, the results will prove very little, except that it is a film you either love or hate – like Marmite!

For example, if market research into soft drinks reveals that there are more consumers of soft drinks in the Midlands than in the South, North, or in London, then 'Midlands' would be the modal category. (Note that it is not possible to calculate a mean or median with category variables.)

Multi-modal distributions

If two or more values are equal first in order of frequency, then the distribution is said to be **multi-modal**. For example, a survey to find out who consumers would respond most positively to in an advertisement for a sports drink yielded the following results:

Anna Kornikova   = 36
Tim Henman   = 36
David Beckham   = 28

Here we have a **bi-modal** result with two favourite categories, Kornikova and Henman, each scoring 36.

The mode always represents a recorded value, unlike the mean which may fall between 2 or more recorded values.

2.4 THE RANGE

Key Ideas

Variation and dispersion

The **range** is the difference between the highest and the lowest values in a data set. It is is useful because it gives a picture of **variation**.

For example, take the case of an examination in which the best student scores 95 and the weakest student scores 15. The range is therefore

95 marks – 15 marks = 80 marks

In this case there is a broad range, which shows that the exam is effective in differentiating between strong and weak candidates. If the results were all bunched between 60 and 65, the range would only be 5, showing little differentiation.

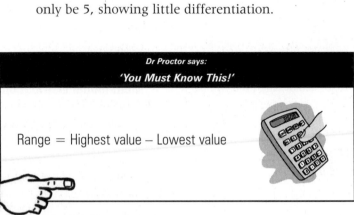

Dr Proctor says:

'You Must Know This!'

Range = Highest value – Lowest value

Doctor Proctor outlines... DISPERSION

In business there are many situations in which it is important to be able to measure the extent of **variation** or **dispersion** around a 'normal' or average value.

For example, take the stock market trading figures published in the financial pages of the daily press. On 18 June 2002, share price fluctuations over the previous 12 months for some leading retailers were as follows:

General Retailers

	High	Low	Current price
Boots	725	585	642
Dixons	262	180	202
Marks & Spencer	423	235	356

Looking at these figures, we can see that each had the following range:

Boots	725 – 585	= **140**
Dixons	262 – 180	= **82**
Marks & Spencer	423 – 235	= **188**

Examining the range makes it possible to identify the extent of dispersion from an average figure. In each of the cases above there is extensive dispersion, indicating the relative volatility of share prices during the period in question.

Investors who are prepared to take risks may be content to live with this high level of volatility. More cautious investors may prefer to put their money into companies which experience smaller fluctuation in share prices – for example, so-called 'blue chip' (i.e. big, well-established) companies.

2.4 THE RANGE

Here are some of the key advantages and disadvantages of the range, with practical examples of each type of use.

Advantages and disadvantages of the range

- Simple, cheap, and easy to collect data.

> **Example**
>
> A bottling company uses automatic equipment to fill bottled drinks. Too large a dispersion in the amount of liquid in the bottles has economic consequences. Too large a spread can result in some bottles being too full, resulting in leakage and waste; under-filling can lead to complaints and possible prosecution.
>
> Every hour the company examines a sample of bottles and checks the dispersion of the filling process. The results have to fall between a minimum and a maximum level. The range must be as small as possible.

- Good for measuring dispersion in a small sample, but less good for large samples.
- Gives the same result from different starting points.

> **Example**
>
> In an exam, one class of students may achieve marks ranging between 70 and 50, and another between 50 and 30. The range is the same – 20, but there is a clear difference between the average performance of each set of students.

- Can be distorted by extreme values.

> **Example**
>
> In a survey aimed at measuring the costs and benefits of a new road scheme, people whose houses are on the route of the road are asked how much money they would be willing to accept in compensation if the road is built.
>
> Answers might lie in the range £100,000 – £500,000.
> But what happens if someone gives an extreme value, e.g. £100 billion or infinity? Clearly this would seriously distort the dispersion of answers.

2.5 QUARTILES AND THE INTERQUARTILE RANGE

Key Ideas

What is a quartile?

In order to use statistics accurately, it is important to avoid distortion.

Sometimes when measuring dispersion around an average, it is necessary to concentrate on figures which lie close to the average and exclude extreme or exceptional values. This can be done by eliminating the **lower** and **upper** **quartiles** of the distribution.

As we saw earlier *(page 26)*, the **median** is the middle value in a distribution. Therefore, half the figures are distributed above the median and half below it.

A **quartile** represents the middle value between two *quarters* of a distribution.

- The **lower quartile** is the value between the first and second quarter of the distribution.
- The **upper quartile** is the value between the third and fourth quarter of the distribution.
- The **interquartile range** is the difference between the upper and lower quartiles.

Dr Proctor says:
'You Must Know This!'

Interquartile range = Upper quartile − Lower quartile

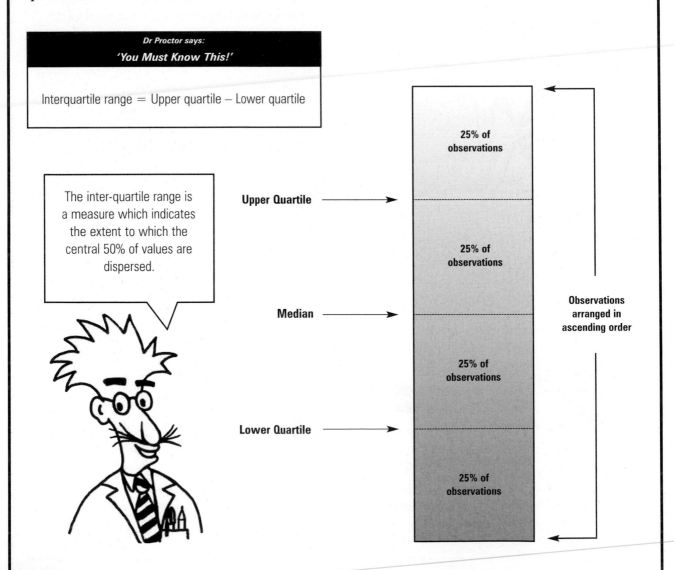

The inter-quartile range is a measure which indicates the extent to which the central 50% of values are dispersed.

Upper Quartile →

Median →

Lower Quartile →

25% of observations

25% of observations

25% of observations

25% of observations

Observations arranged in ascending order

2.5 QUARTILES AND THE INTERQUARTILE RANGE

Doctor Proctor outlines... CALCULATING THE INTERQUARTILE RANGE

The figures below show an array of 40 prices which different booksellers are charging for the same book – some over the Internet and others through wholesale outlets.

£

18	20	20	23	24	24	25	25	26	26
27	27	27	28	28	28	28	29	29	29
30	30	30	30	30	31	31	31	31	32
32	35	35	35	36	39	39	39	40	42

The total number of observations (*n*) is 40.

- The **lower quartile** can be calculated by taking away the bottom quarter of observations – i.e. the first ten numbers. The lower quartile will therefore be the number between 26 (the 10th number) and 27 (the 11th number) = 26.5.

- The **median** lies between the lower 50% of numbers and the top 50%. As the 10th number is 29, and the 30th is 30, the median is 29.5.

- The **upper quartile** can be calculated by taking away the top quarter of observations – i.e. the last ten numbers. The upper quartile will therefore be a number that lies between 32 (the 30th number) and 32 (the 31st number) i.e. 32.

- The **interquartile range** is therefore 32 – 26.5.

Deciles and percentiles
In addition to breaking down figures into quartiles, it is possible to break them down further into **deciles** (tenths) and **percentiles** (hundredths).

For example, in comparisons of income and earnings, the population is often broken down into percentiles and deciles. This allows economists to compare the earnings of the top and bottom 10% of the population.

The interquartile range often provides a better measure of dispersion than the full range. This is because it eliminates extreme values which can distort the overall picture.

2.6 THE STANDARD DEVIATION

Key Ideas 🔑

Deviation: a measure of dispersion

Standard deviation is the most commonly used measure of dispersion. Like the mean, it takes into account every observation in a data set.

The table below shows the sales revenue of a business over a period of five weeks.

	Week 1	Week 2	Week 3	Week 4	Week 5
Sales Revenue (£)	100,000	100,000	100,000	100,000	100,000

In this case, the mean revenue is clearly £100,000, and there is no deviation from this figure. However, in most situations there *will* be some deviation from the mean. We use the term **standard deviation** to describe the 'average' deviation from the mean.

Doctor Proctor outlines... CALCULATING STANDARD DEVIATION

The following table shows the number of faults reported on two production lines in a factory over a weekly period:

	Mon	Tue	Wed	Thu	Fri	Sat
Line 1	20	22	24	22	24	20
Line 2	15	30	60	15	20	40

It is obvious that the figures for Line 2 are much more dispersed than those for Line 1, so we can expect the standard deviation to be larger.

Let's look first at the deviation for Line 1.

The **mean** for Line 1 is $20 + 22 + 24 + 22 + 24 + 20 = 132 \div 6 = 22$. The deviations from the mean are therefore:

	Mon	Tue	Wed	Thu	Fri	Sat
Value	20	22	24	22	24	20
Deviation	−2	0	+2	0	+2	−2

The greater the dispersion of observations from the mean, the bigger will be the deviations and thus the greater the standard ('average') deviation.

The next step is to calculate the standard deviation. There would be no point in taking an arithmetic mean of the deviations as these would always add up to zero (the positive deviations will exactly cancel out the negative ones). What we need to do is to square the deviations in order to get rid of the minus signs:

Value:	20	22	24	22	24	20
Deviation from 22:	4	0	4	0	4	4

We use the term **variance** to describe the mean of the squared deviations. In this case the variance is:

$$\frac{4 + 0 + 4 + 0 + 4 + 4}{6} = \frac{16}{6} = 2.66$$

contd on page 33

2.6 THE STANDARD DEVIATION

Doctor Proctor outlines... CALCULATING STANDARD DEVIATION (CONTD)

contd from page 32

The problem of using variances is that they are set out in the form of the square of an observation. Often this has very little practical use. For example, if we are measuring the level of absenteeism in a factory, this might be set out in terms of days per week or per month. When we examine the variance this would appear as 'squared days per week' – a difficult concept to understand!

We therefore need to convert our figures back into more meaningful terms, i.e. the original observed values. This involves calculating the square root of the variance. The standard deviation for Line 1 is thus $\sqrt{2.66}$.

Now let's work out the standard deviation for Line 2.

Referring back to the table, we can see that the mean value for Line 2 is $15 + 30 + 60 + 15 + 20 + 40 = 180 \div 6 = 30$.

Value	15	30	60	15	20	40
Deviation from mean (30)	−15	0	+30	−15	−10	+10
Squared deviation	225	0	900	225	100	100

$$\text{Variance} = \frac{225 + 0 + 900 + 225 + 100 + 100}{6} = \frac{1550}{6}$$

$$= 258.33$$

Standard deviation for Line 2 is therefore $\sqrt{258.33} = 16.07$.

It can be seen from this that the standard deviation for Line 2 is greater than for Line 1.

Dr Proctor says:
'You Must Know This!'

Standard deviation is the square root of the variance.

The standard deviation for Line 1 is therefore:

$\sqrt{2.66} = 1.63$

Uses of standard deviation

Standard deviation takes into account all the observations in a population or sample and is a useful measure of dispersion. It can be used for:

- quality control or checking variations against a standard or average
- identifying variations in sales or production figures
- establishing acceptable standard deviations for use as benchmarks. Where figures exceed the acceptable standard deviation, controlling action can be taken to remedy the situation.

However, when data are skewed, i.e. when the mean is towards one or other end of the range, perhaps as a result of some exceptional observations, it is better to use the interquartile range.

2.1 The mean

1   The figures opposite were taken from a market research survey showing the distance travelled by customers to a shop. Calculate:
    a   The mean distance travelled by shoppers
    b   i) The mean distance travelled by shoppers with cars; ii) the mean distance travelled by customers using the bus
    c   The mean distance travelled by i) male shoppers; and ii) female shoppers.
2   What are the advantages of the mean as a way of working out averages?

2.2 The median

In a customer satisfaction survey, shoppers were asked to rank the service in a café as follows:

    i   Excellent
    ii   Very good
    iii   Good
    iv   Average
    v   Poor

The results of a sample of 50 customers were as follows.

Customer	Response	Customer	Response	Customer	Response
1	iii	18	i	35	i
2	iii	19	i	36	i
3	iii	20	i	37	ii
4	i	21	i	38	i
5	ii	22	i	39	ii
6	iv	23	ii	40	i
7	ii	24	i	41	i
8	v	25	ii	42	i
9	i	26	ii	43	ii
10	i	27	ii	44	ii
11	ii	28	v	45	ii
12	ii	29	i	46	i
13	v	30	i	47	iii
14	iv	31	i	48	ii
15	ii	32	ii	49	i
16	i	33	ii	50	i
17	i	34	ii		

1   What is the median response given by consumers? Show your working.
2   What advantages are there to using the median as a way of averaging figures?
3   Why might the median earnings of employees provide a better average than the mean?

Shopper reference no.	Sex Male (1) Female (2)	Method of transport Car (1) Bus (2) Foot (3)	Distance travelled (miles)
1	1	1	40
2	1	2	10
3	2	3	5
4	2	1	25
5	2	3	3
6	1	2	8
7	1	1	8
8	2	1	12
9	1	2	8
10	1	1	50
11	2	3	1
12	2	1	64
13	2	2	12
14	1	1	2
15	2	3	10
16	2	2	20
17	1	2	25
18	1	1	87
19	1	1	30
20	2	2	2
21	2	3	1
22	2	1	45
23	1	3	3
24	1	2	5
25	1	1	60
26	2	3	1
27	2	3	1
28	2	2	10
29	2	1	14
30	2	2	6
31	2	3	2
32	1	3	3
33	1	2	12
34	2	1	2
35	1	1	6
36	2	2	8
37	2	3	1
38	1	1	30
39	2	1	60
40	1	3	1

2.3 The mode

In April 2000 a Sheffield United supporter claimed that Sheffield Wednesday's 'average' number of goals scored during the season at home was 'a big fat zero'. He made this assertion on the basis of the evidence below.

Note: All the matches shown are home matches. Sheffield Wednesday goals for home matches appear first.

v. Arsenal	1–1
v. Bradford	2–0
v. Coventry	0–0
v. Derby	0–2
v. Everton	0–2
v. Liverpool	1–2
v. Manchester United	0–1
v. Middlesborough	1–0
v. Newcastle	0–2
v. Southampton	0 1
v. Tottenham	1–2
v. Watford	2–2
v. West Ham	3–1
v. Wimbledon	5–1

1 What was the basis for the Sheffield United fan's average?
2 How else might the 'average' have been interpreted?
3 What do you think would have been the best way of calculating the average?

2.4 The range

1 What is meant by the **range** of a set of values?
2 How useful is the range as a measure of dispersion?
3 What problems are caused by the existence of extreme values when measuring dispersion?

2.5 The quartile and interquartile range

1 What is meant by the following:

  a Upper quartile
  b Lower quartile
  c Interquartile range
  d Decile
  e Percentile

2 Why does the interquartile range give a better picture of dispersion than a simple range?

2.6 Standard deviation

Two production lines have been set up to package 34 bite-size sweets into bags. Regular quality checks are made to identify the number of sweets in each finished bag. The results of the most recent sample checking are:

Line 1	34	34	35	38	32	33	33	34	33	34
	37	32	34	34	34	32	33	34	40	23
	34	34	34	33	33	31	30	32	32	24
	32	49	32	32	30	30	30	34	34	34
Line 2	34	33	34	35	35	33	34	33	33	33
	52	34	34	34	34	34	32	34	34	34
	34	33	33	34	34	34	35	35	35	32
	33	34	34	35	35	32	34	34	34	34

1 Calculate the range for each of the two production lines.
2 Calculate the interquartile range for each of the two production lines.
3 Calculate the standard deviation for each of the two production lines.
4 What are the main advantages and disadvantages of standard deviation as a measure of dispersion?

UNIT 3

PROBABILITY

Topics covered in this unit

3.1 The probability of events occurring
Defining probability; calculating probability on the basis of possible outcomes and past experience.

3.2 The probability of combinations of events
Calculating the probability of combinations of events which may or may not be linked.

3.3 Expected values
Calculating the mean of a probability distribution (its **expected value**); the importance of expected value in managing uncertainty.

3.4 Decision trees
How to construct and use decision trees to trace through the possible outcomes of a decision.

3.5 The normal distribution
A **normal distribution** is symmetrical around the mean.

3.6 Sampling
The importance of random selection in sampling; statistical checks to ensure random sampling.

The concept of **probability** is relevant to a wide range of business decision-making. Calculating probabilities enables businesses to predict the likelihood of events taking place or outcomes occurring. It also helps managers to assess risk and provides evidence to support taking, or not taking, a particular course of action.

Probability is especially important for businesses which assign a monetary value to certain types of risk — for example, insurance companies, investment analysts, companies about to market a new product, etc.

3.1 THE PROBABILITY OF EVENTS OCCURRING

Key Ideas 🔑

The importance of probability

There are many situations in which business managers may need to calculate the **probability** of an event or outcome.

For example, in a competitive market, a manufacturer may need to assess the likelihood of winning an important contract. Here, the odds can be calculated reasonably accurately, based on past experience and on knowledge of the market.

However, some future events can be much harder to predict because they involve new situations or factors which are complex and unforeseeable.

Macro-economic events such as the recession in world trade following the attacks on the World Trade Center in New York on September 11 2001 make business planning very difficult.

Major economic events such as Britain joining the Euro or a short-term rise or fall in interest rates are hard to predict because they involve political as well as economic factors.

In these cases, calculating probabilities can involve weighing up many different issues and factors.

Doctor Proctor outlines... DEFINING PROBABILITY (1)

Probability based on number of possible outcomes

One definition of probability is 'the extent to which an event is likely to occur, measured by the ratio of the favourable cases to the whole number of cases possible'.

For example, in the case of someone trying to throw a 'double' in a board game, there are 36 possible outcomes:

(**1,1**)	(1,2)	(1,3)	(1,4)	(1,5)	(1,6)
(2,1)	(**2,2**)	(2,3)	(2,4)	(2,5)	(2,6)
(3,1)	(3,2)	(**3,3**)	(3,4)	(3,5)	(3,6)
(4,1)	(4,2)	(4,3)	(**4,4**)	(4,5)	(4,6)
(5,1)	(5,2)	(5,3)	(5,4)	(**5,5**)	(5,6)
(6,1)	(6,2)	(6,3)	(6,4)	(6,5)	(**6,6**)

Of the 36 possibilities, only the six shown in bold represent doubles. The probability of throwing a double is thus 6/36 or 1/6.

In words, this ratio can be expressed as:

$$\frac{\text{The number of ways an event of interest can occur}}{\text{The total number of equally likely outcomes}}$$

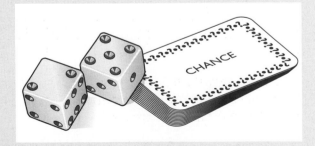

3.1 THE PROBABILITY OF EVENTS OCCURRING

Doctor Proctor outlines... DEFINING PROBABILITY (2)

Probability based on past experience

In calculating probabilites, business managers nearly always need to take acccount, not only of the total number of possible outcomes, but of past experience.

This is why, before launching a new product, businesses will often carry out trials on a small sample of the market. Using these results, it is possible to calculate the probability of success with the population as a whole.

Setting out a probability distribution

In order to calculate probabilities in this way, it is first necessary to set out a table showing the relative frequency of the different possible outcomes.

By analysing the relative frequency of past events, we can calculate the probability of their occurring in the future – always assuming that the past is a representative guide to future probabilities.

The example below shows how a business might estimate the number of customers using its website at certain periods.

Column 1 shows the absolute frequency of past visits.

Column 2 shows the relation of actual number of visits to total number of observations.

Column 3 shows the probability of the same patterns being repeated in future.

Example

The Internet booksellers **BooksUnlimited.com** wish to measure the probability of large numbers of customers logging on to their website at certain peak periods.

To gauge future demand, the company records the number of visits made by customers over a total of 100 15-minute periods, all at peak times of the day. The results are shown in the table below.

Visits Made by Customers to BooksUnlimited. com Website

1 Absolute frequency distribution		2 Relative frequency distribution		3 Probability distribution	
Variable	Frequency	Variable	Relative frequency	Variable	Probability
0–5	10	0–5	10/100	0–5	0.10
6–10	15	6–10	15/100	6–10	0.15
11–20	15	11–20	15/100	11–20	0.15
21–30	10	21–30	10/100	21–30	0.10
31–40	20	31–40	20/100	31–40	0.20
41–50	30	41–50	30/100	41–50	0.30

The table shows that 0–5 customers logged on to the site 10 times (absolute frequency) making up one-tenth of the observations (relative frequency). This gives a probability of 0.10 of the same thing happening in future peak periods.

3.1 THE PROBABILITY OF EVENTS OCCURRING

Key Ideas 🔑

Probability distribution

As we saw on page 39, in some situations future probabilities can be calculated on the basis of the past frequency of different outcomes. These outcomes need to be:

- exhaustive (i.e. including all possible outcomes)
- mutually exclusive *(see example right)*.

However, the problem with many decisions in business is that they involve new situations where outcomes are hard to predict.

Managers therefore have to base their calculations at least partly on 'hunches' and guesswork. For example, in launching a new product, a marketing manager may say: 'Based on evidence from similar product launches in similar markets, there is a 50% probability of the product making substantial profits'.

Example

Possible errors in a quality control exercise are grouped in the following categories:

- 0 to less than 0.3
- 0.3 to less than 0.6
- 0.6 to less than 0.9

An error level of 0.2 can therefore *only* be recorded in the category '0 to less than 0.3'. All recordings of less than 0.3 are recorded in the category '0 to less than 0.3'.

Testing over time has shown the following probabilities of different-size errors occurring:

Variable (X) (size of error)	Probability
0 to less than 0.3	5/10
0.3 to less than 0.6	3/10
0.6 to less than 0.9	2/10
Total	1

Dr Proctor says:
'You Must Know This!'

The 'relative frequency' definition of probability
The 'relative frequency' definition of probability states that, if in a number of trials (n), r of these trials result in event E, the probability (P) of event E is r/n.

The formula for probability can therefore be expressed as:

$$P(E) = \frac{r}{n}$$

Probabilities are generally expressed in terms of values between 0 and 1. So a probability of 0.5 represents a 5 in 10 chance of an event happening.

If something is *certain* to happen, i.e. if $r = n$, the probability is 1.

Probabilities can be converted to percentages by multiplying by 100. So a probability of 5/10 represents a percentage of 50%.

3.2 THE PROBABILITY OF COMBINATIONS OF EVENTS OCCURRING

Key Ideas

The multiplication ('and') law of probability

In business and economics companies often need to calculate the probability of more than one event occurring at the same time.

For example, a business may be encouraged to invest if *both* the tax rate *and* the interest rate are falling.

In order to calculate the probability of both events occurring – or the probability of P(E1 and E2) – we need to apply the **multiplication law of probability**.

Doctor Proctor outlines... CALCULATING MULTIPLE PROBABILITIES

In simple terms, the multiplication ('and') law of probability involves multiplying one probability by the other.

For example, suppose that a business wants to invest in a new head office, but will only do so if interest rates are below 5%, *and* the level of corporation tax is reduced by at least 2%. It therefore needs to work out the probability of these two events occurring together.

Assume that the probability of interest rates being below 5% is 4/10, and the probability of corporation tax being reduced by 2% and over is 2/10:

The probability of these two events occurring at the same time is

$$\frac{4}{10} \times \frac{2}{10} = \frac{8}{100} = 0.08$$

In this case the two events E1 (fall in interest rates below 5%) and E2 (fall in corporation tax by at least 2%) do not depend on each other, so we say that they are **statistically independent**.

Note that the probability of the two events occurring together (8/100) is much smaller than the probability of just one of the events occurring (4/10 and 2/10 respectively).

The probability of a single event occuring is normally much higher than the probability of that event combining with one or more other events.

3.2 THE PROBABILITY OF COMBINATIONS OF EVENTS OCCURRING

Key Ideas 🔑

Linked probabilities

In the previous section we looked at calculating probabilities which are statistically independent.

In many cases, however, businesses need to calculate the probability of events which are combined or linked in some way.

In order to do this, it is necessary to apply the **addition law of probability**.

> In the real world, businesses need to consider the chance of several events combining — i.e. linked probabilities.

Example

The table below shows the probability of a firm's sales revenues falling within certain financial limits or bands over a given period.

Event	Probability
(Sales revenue)	
£0 – £20,000	1/10
£20,001 – £30,000	2/10
£30,001 – £40,000	5/10
£40,001 – £50,000	2/10

Clearly, the most desirable situations from the firm's point of view lie at the top end of the range.

To find out the probability of sales being greater than £30,000 but less than £50,000, we can add the probabilities of the top two events, i.e.:

5/10 + 2/10 = 7/10

So the probability of sales being between £30,001 and £40,000 **and** between £40,001 and £50,000 is 7/10.

3.3 EXPECTED VALUES

Key Ideas 🔑

Estimating financial outcomes

Business often have to make difficult choices.

Choosing between Project **A** which will make a profit of £100,000 and Project **B** which make a profit of only £20,000 is clearly *not* difficult. However, when probability is involved, decisions become more complex.

Suppose for example that the probability of Project A succeeding is only 1 in 10, whereas the probability of success for Project B is 7 in 10. Which would you choose?

Here, managers need to have a means of calculating not just the financial outcome of a decision but also its probability. The term **expected value** describes this balance of profit and loss and probability. It is based on the predicted profit or loss of an event *and* the probability of that event occurring.

Example

A manufacturer has carried out detailed research into consumer opinion about a new hazelnut-flavoured chocolate bar.

The research indicates that there is a 1 in 5 chance of the hazelnut chocolate bar being a success when launched on the market. If the bar is successful, a £20-million profit is expected. However, there is a 4 in 5 chance that the hazelnut chocolate bar will fail. If this is the case, there will be a loss of £2 million.

We can therefore work out the expected value of launching the new chocolate bar. This is:

Expected value =

Success			**Failure**		
0.2	x	£20m	+	0.8	x	—£2m
(Probability)	(Expected profit)		(Probability)	(Expected loss)	

= £4m – £1.6m

= £2.6m

An alternative decision available to the firm is to make a strawberry-flavoured chocolate bar. Here, research indicates that there is a 50% chance that the bar will be a success and that this will bring in profits of £10 million. However, if the bar fails (a 50% chance) then losses will be £4 million.

We can illustrate the alternative (strawberry-flavoured) decision in the following way:

Expected value =

Success			**Failure**		
0.5	x	£10m	+	0.5	x	—£4m
(Probability)	(Expected profit)		(Probability)	(Expected loss)	

= £5m – £2m

= £3m

Comparing chocolate with strawberry, we can see that it makes more sense to go ahead with the strawberry-flavoured bar because it is more likely to bring in a higher profit.

> Estimating financial outcomes involves linking probabilities with expected values.

3.4 DECISION TREES

Decision trees are a simple graphic way of choosing from alternative courses of action when faced by uncertainty.

For example, an investor (or investment trust) might set out a decision tree enabling them to balance the twin objectives of growth and security.

The basic procedure for constructing a decision tree is to set out a series of alternative decisions as branches of the tree and then to calculate the probability of the success of the event and the likely financial return.

A worked example is shown below and on pages 45–48.

Look at the three options set out here.

Which course of action would YOU choose?

Now compare your answer with the outcome shown on page 48.

Example

As a result of a take-over, a business has acquired an empty factory and is now faced with the following choices:

- **Option 1:** Sell the factory immediately for £125,000

- **Option 2:** Hold it for a year in the hope that property prices will rise and then sell it for £400,000. There is a 1/10 probability that they will be able to sell it at this price in a year's time. However, there is a 9/10 chance that property prices will drop. This will mean the sale of the factory will raise only £100,000

- **Option 3:** Refit the factory and use it for manufacturing. This will cost £200,000.

If the business takes Option 3 above, it will have three further choices:

a   Manufacture a new product – but there is a 50% chance that this will make no profit at all

b   Extend production of an existing popular product – with a 40% chance of making a £400,000 profit

c   Produce a modified product – with a 10% chance of making a profit of £1,500,000.

3.4 DECISION TREES

Example (contd)

The three alternatives under Option 3 can be set out in a table as follows:

Possible Outcomes from Developing the Factory Now

Type of production	Probability	Expected profit (£)
New product	5/10 (0.5)	0
Existing popular product	4/10 (0.4)	400,000
Modified product	1/10 (0.1)	1,500,00

This information forms the basis of the decision tree *(below)*.

Pages 46–48 take you step by step through the process by which it has been constructed.

> **Dr Proctor says:**
> **'You Must Know This!'**
>
> In a decision tree, it is possible to distinguish between points of decision and points where chance and probability may come into play.
>
> - **Decision forks** are represented by squares
> - **Chance forks** are represented by circles.
>
> In the example here, the decision at the fork is to sell the factory now, sell later, or to use the factory for manufacturing. Note that the probability lines all come out of the circles.

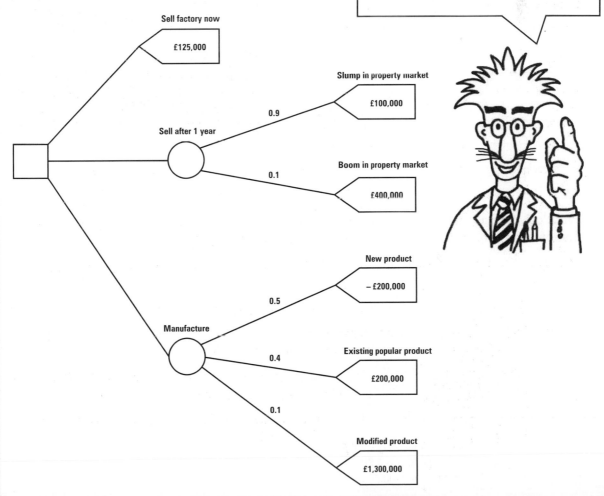

3.4 DECISION TREES

Key Ideas 🔑

Constructing a decision tree

On page 48 you can see the completed decision tree showing the three options the company can choose from in our example:

- sell the factory now
- sell later
- keep the factory and refit it for manufacturing.

Now let's see how the decision tree was constructed.

Example (contd)

STEPS TAKEN IN CONSTRUCTING THE TREE

1 **Set out the decision fork**, i.e. the square on the left-hand side of the diagram. Three lines drawn from it represent the three possible alternatives – sell now, sell later, or manufacture.

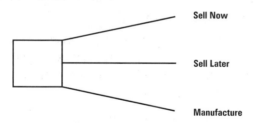

Sell Now

Sell Later

Manufacture

2 **Work out the consequences of each course of action.**

- The consequence of **Option 1** is the easiest to set out. There is only one possible outcome: the proceeds of £125,000 from the sale of the factory.
- **Option 2** yields alternative outcomes because of the chance element. These are shown as lines emanating from a circle. There are two possible outcomes and these are labelled with their probabilities (0.90 and 0.10).

How do we know what the financial effect of these outcomes will be? Selling the factory next year has a 9/10 chance of yielding only

£100,000 and a 1/10 chance of yielding £400,000. (Note that in this example we have ignored the fact that money earned in the future is worth less than money held now).

- **Option 3** involves setting out a circle to represent an even more complicated chance fork. In each of the three cases we need to subtract £200,000 (the cost of refitting the factory).

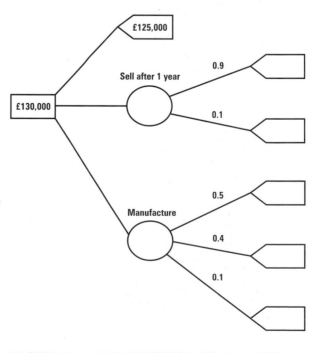

£125,000

0.9

Sell after 1 year

£130,000

0.1

0.5

Manufacture

0.4

0.1

3.4 DECISION TREES

Example (contd)

The purpose of the decision tree is to help you to choose the best course of action in a given set of circumstances.

In order to do this, you need to work back from right to left along the decision tree, filling in each chance and decision fork (at present they are just empty circles and squares).

Each chance fork needs to be averaged, and each decision fork 'pruned back'.

STEPS TAKEN IN CONSTRUCTING THE TREE (CONTD)

3 Fill in the chance forks and decision forks.

- **Option 1** – to sell now – needs no alteration, because it contains only one figure: £125,000.

- **Option 2** – to sell in a year's time – needs to be averaged out. We must take the mean (average) value or **expected value (EV)** *(see page 43)* by **weighting** (multiplying) each outcome against its probability.

The expected value of selling the factory next year is therefore:

$$EV = 0.9\,(100{,}000) + 0.1\,(400{,}000)$$
$$= 130{,}000$$

This figure of 130,000 can now be added to **Option 2** chance fork (in the circle).

- **Option 3** – using the factory for manufacturing – now needs to be calculated (in each option, I have deducted £200,000, which is the cost of refitting the factory).

The expected value of manufacturing is:

$$0.5\,(-200{,}000) + 0.4\,(200{,}000) + 0.1\,(1{,}300{,}000)$$
$$= 110{,}000.$$

We can now fill in the 110,00 in the **Option 3** chance fork (in the circle).

3.4 DECISION TREES

Example (contd)

'PRUNING BACK' THE DECISION TREE

Once the decision forks and chance forks on the decision tree have been filled in, the final step is to 'prune it back' – in other words, to eliminate the less attractive options so as to leave the best course of action.

This is done by comparing the expected values of the three options:

- Expected value of selling factory now = £125,000
- Expected value of selling in one year = £130,000
- Expected value of manufacturing = £110,000

Having cut off the less attractive branches, we are left with just one expected value in the decision square, i.e. £130,000.

The diagram (*below*) shows how the final decision has been made.

It is important to remember that, in creating decision trees, organisations are often concerned with very large-scale probabilities.

For example, when a major retailer is faced with a choice of either redesigning the layout of its stores or leaving the stores as they are, hundreds of separate outlets may be affected.

In effect, the retailer may be working out a probability that the change will raise revenues by a given sum in, for example, 60% of stores.

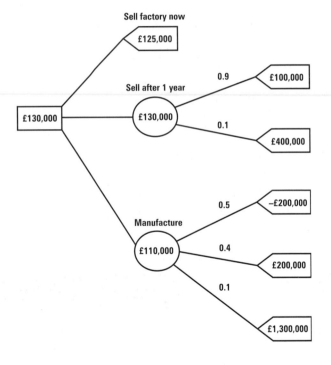

Solving the decision tree (averaging the chance forks and pruning back the decision forks)

3.5 THE NORMAL DISTRIBUTION

Key Ideas 🔑

Distribution curves

One of the simplest ways to illustrate the frequency of observations is to draw a **distribution curve**. This is done by joining together the mid-points of bars in a histogram.

For example, the histogram and table *(right)* relate to the length of time (in seconds) it takes viewers to recognise a particular make of car in a TV commercial. The observations are made in relation to class intervals of 5 seconds each.

Time taken to recognise brand name in TV car commercial	
Class interval	*Frequency of observations*
0–5 seconds:	4
6–10 seconds:	10
11–15 seconds	18
16–20 seconds	13
21–25 seconds	11
26–30 seconds	8
31–35 seconds	5
36–40 seconds	3
41–45 seconds	1

You can see that in both the histogram and the corresponding distribution curve *(right)*, the long tail of observations lies on the right-hand side of the diagram.

This type of distribution is said to be **positively skewed**. In a **positively skewed** distribution the mean, median and mode lie on the left-hand side of the distribution curve.

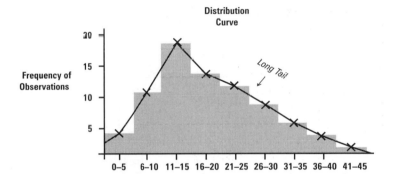

Distribution Curve

Frequency of Observations

Long Tail

0–5 6–10 11–15 16–20 21–25 26–30 31–35 36–40 41–45

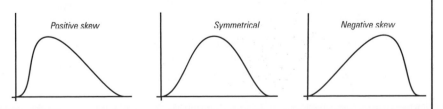

Dr Proctor says:

'You Must Know This!'

- In a **positively skewed** distribution curve, the mean, median and mode lie on the left and the tail on the right.
- In a **symmetrical** distribution curve the mean, median and mode lie in the centre.
- In a **negatively skewed** distribution curve, the mean, median and mode lie on the right of the diagram and the tail is on the left.

Positive skew   *Symmetrical*   *Negative skew*

3.5 THE NORMAL DISTRIBUTION

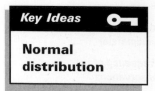

Key Ideas

Normal distribution

One of the most common distributions is the **normal distribution**.

The use of the term 'normal' does not mean that the distribution is necessarily the 'usual' curve – rather, that it is an 'idealised' curve, against which we can make comparisons with observations in the real world.

In a normal distribution curve, the majority of observations are close to the mean, and relatively few observations are further away from it.

For example, in an industry in which many very similar firms compete with one another, we would expect most firms to make profits which are close to the mean for the industry and relatively few to make much higher or lower profits than the rest.

The more observations that we plot on a normal distribution curve, the clearer its characteristic 'bell shape' becomes.

Note the bell shape of a normal distribution curve. This is because the highest frequency of observations will be at, or close to, the mean, which lies at the centre of the distribution curve.

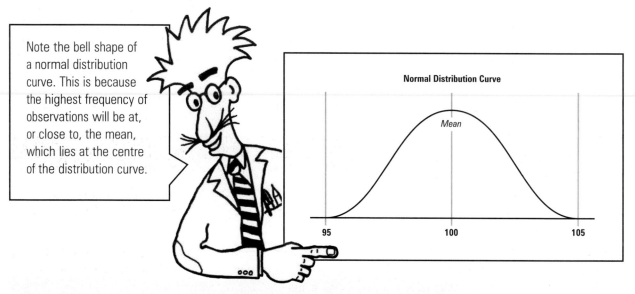

Normal Distribution Curve

Mean

95   100   105

Doctor Proctor outlines... CHARACTERISTICS OF NORMAL DISTRIBUTION

1   The shape of a normal distribution is completely determined by the position of its mean and its standard deviation. The mean determines where the curve is located: the centre of the curve will be at the value of the mean on the horizontal axis.

For example, if the mean amount of money spent by consumers in a supermarket is £50, then the curve will be drawn around a centre

of £50. The standard deviation of the curve will determine its dispersion to the left and right of the centre.

2   The normal distribution has a symmetrical shape around the mean and is represented by a bell-shaped curve.

3   In a normal distribution the mean, the median and the mode coincide.

contd on page 51

3.5 THE NORMAL DISTRIBUTION

Doctor Proctor outlines... CHARACTERISTICS OF NORMAL DISTRIBUTION (CONTD)

4 The variable **x** extends from minus infinity to plus infinity. The relative frequency of large (plus or minus) values of **x** is very small and can be ignored.

5 Although the normal distribution curve is always symmetrical, it can be tall and thin, short and stocky, or quite flat. The shape depends on the size of the standard deviation – i.e. whether it is large or small.

Another factor determining what the curve looks like is the scale chosen for the vertical and the horizontal axis of the graph *(below)*.

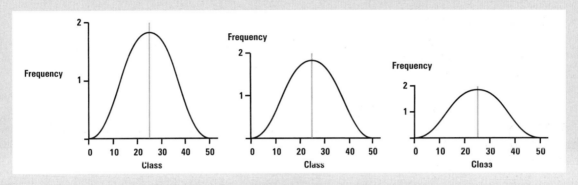

The normal distribution curve has certain mathematical properties which enable us to calculate precisely what percentage of the population will lie between any two values of the variable.

So long as we know the mean and the standard deviation of a distribution curve, we can regard any particular value in the distribution as being 'so many standard deviations away from the mean'.

Dr Proctor says:
'You Must Know This!'

In a normal distribution curve, 50% of observations are equal to or greater than the mean and 50% are equal to or lower than the mean.

About two-thirds (68%) of all observations lie within **one standard deviation** either side of the mean (i.e. one-third above and one-third below).

Example
If we know that the mean profit in a competitive industry is £50,000 and that the standard deviation is £10,000, a firm whose profit is £60,000 can be described as being 'one standard deviation above the mean'.

We can show examples of these standard deviations in a table as follows:

Profit	Number of Standard Deviations from Mean
£50,000	Mean value
£60,000	1 standard deviation above the mean
£40,000	1 standard deviation below the mean
£70,000	2 standard deviations above the mean
£30,000	2 standard deviations below the mean

3.5 THE NORMAL DISTRIBUTION

> ### Key Ideas 🔑
>
> ### 'Slicing up' the normal distribution curve

The bell-shaped normal distribution curve can be split up into standard slices, each one containing a known percentage of the total observations.

A detailed **normal distribution table** (not included in this book) can be used to calculate the area from the mean to a fixed number of standard deviations away from the mean.

The letter Z represents the number of standard deviations a value is above or below the mean. It is calculated by:

$$Z = \frac{x - m}{s}$$

Where:

$Z =$ number of standard deviations above or below the mean

$x =$ the value

$m =$ the mean

$s =$ the standard deviation

Doctor Proctor outlines... THE POINT OF INFLECTION

As we have already seen, there is a characteristic shape to the normal distribution curve. Starting at the **apex** or highest point, of the curve, it begins by falling outwards in a convex shape. However, at a certain point it starts to bend back and to bulge inwards in a concave shape. The point at which the shape of the curve changes from being convex to concave is termed the **point of inflection**.

The distance on the curve between the mean (at the centre) and the point of inflection is that of one standard deviation. About two-thirds of all observations fit within one standard deviation of the mean.

If 68% of observation lie within one standard deviation of the mean, common sense tells us that the remaining 32% of observations are greater than one standard deviation from the mean.

> **Dr Proctor says:**
> ### 'You Must Know This!'
>
> **68%** of all observations lie within one standard deviation from the mean.
>
> **95%** of all observations lie within two standard deviations from the mean.

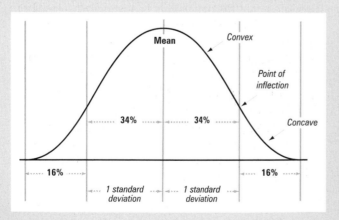

In a normal distribution curve, 68% of all observations are within one standard deviation from the mean

3.6 SAMPLING

Key Ideas 🔑

Samples and Total Populations

In business and economics investigations, researchers often examine samples rather than entire populations.

- in market research, several hundred people may be selected from a total population which may contain millions.
- In quality control checking and inspection, it is also usual to examine a sample of items for possible faults or imperfections.

The larger the size of the sample, the more representative it becomes of the total population – and thus the more likely it is to be close to a normal distribution.

Information collected by surveying every individual in a population, such as that obtained in the ten-yearly **Population Census**, comes as close to 100% reliability as possible. However, complete coverage is out of the question in most investigations because of the cost and effort involved.

Dr Proctor says:
'You Must Know This!'

The larger the size of the sample, the more representative it becomes of the total population – and thus the more likely it is to be close to a normal distribution.

In most cases – and this includes the collection of statistics by government departments – only a sample of individuals from a population are selected.

For example, the annual **Family Expenditure Survey** aims to obtain reliable data on household expenditure, income and other aspects of household finances for the whole UK population of 24 million households. However, only some 10,000 households take part in the survey, and of these, only about 7,000 actually cooperate in providing the information requested.

Doctor Proctor outlines... MINIMISING SAMPLING ERROR

Any sample chosen must be as representative of the total population as possible. It must also be chosen in such a way as to minimise the risk of **sampling error**.

The sampling error of a statistic such as average income is the difference between the average income of the sample and the average income for the whole population. Some samples will have only a small sampling error. They will produce an estimate close to the population average. Other samples will show a larger sampling error.

Designing the sample

Accuracy of sampling depends very much on the design of the sampling process. Even though the sample size used to collect data for the Family Expenditure Survey is small in relation to the total number of households (7,000 out of 24 million), efficient sample design enabled average weekly household expenditure for the year 1999/2000 to be estimated with 95% probability at £359. The error each way is as little as 3%, or plus or minus £10.

3.6 SAMPLING

Key Ideas

The importance of random selection

To gain a representative sample you need to select observations randomly. Every individual in the population or **sampling frame** should have an equal chance of being selected.

Try this. Tap out ten numbers between 000 and 999 using the random number function on your calculator. Average out the numbers. Repeat the exercise five times.

Each time, work out the average of the combined samples – 1 sample's average, 2 samples' combined average, 3 samples' combined average, 4 samples' combined average, 5 samples' combined average. You are likely to find that the average of your combined samples moves closer to the mean of the population (499.5).

You should now be able to understand why a sample size of 7,000 used in the Family Expenditure Survey *(see page 53)* gives a very close estimate of the population's mean household expenditure.

Example

A good example of random selection is the National Lottery, where numbers are selected in a random order from balls bouncing down a chute. This gives a known probability of any particular number selection being drawn. The winning numbers are thus a random sample from all possible number combinations.

Ensuring random selection

To ensure that the Lottery draw is fair, it is regularly put to a number of statistical tests to check that the draw is random and unbiased. These are similar to the type of check that can be carried out to ensure that a calculator's 'random number' function is working accurately *(see left)*.

To check whether the 1,000 numbers between 000 and 999 are being generated randomly using the random number function, we can proceed by drawing samples of ten numbers and finding their average. We know that the average of this population is 499.5. The average of a truly random sample of 10 numbers will have a value lying within plus or minus 100 from the population average with a high degree of probability – about 99%. If we take several sample averages and these fall outside these limits, we can be pretty certain that the random number function on our calculator is not working correctly.

The National Lottery: how random is random?

3.1 The probability of events occurring

1 In a random sample, consumers who had just visited a supermarket were found to have the following numbers of grocery items in their trollies:

26   26   26   29   23   24   28   25   24   27

  a   What is the probability of a consumer having exactly 24 items in their trolly?

  b   What is the probability of a consumer having fewer than 24 items in their trolly?

  c   What is the probability of a consumer having *either* 24 or 25 items in their trolley?

2 The number of items on each of a number of shelves in a supermarket is shown in the following table:

Number of items per shelf	Number of shelves
19	2
20	3
21	7
22	5
23	14
24	11
25	12
26	9
27	6
28	6
29	3
30	2

  a   What is the probability of a shelf having exactly 24 items?

  b   What is the probability of a shelf having less than 24 items?

  c   What is the probability of a shelf having either 24 or 25 items?

3.2 The probability of combinations of events occurring

An investor is not sure whether to hold on to her investments in a leisure company. She will do so if she thinks there is a good chance of her shares in the company rising in price *and* the company paying out a good dividend. She knows that there is a 1/5 probability that dividends will be high when the company announces its next results, and that there is a 1/3 probability that the price of the shares will rise.

What is the probability of *both* the company giving a good dividend *and* share prices rising?

3.3 Expected values

1 What do you understand by the term **expected values**?

2 A company is considering introducing a new product to the market. Research evidence indicates that if it launches the new product there is a 2/5 chance that the product will be a success, yielding profits of £4 million. However, there is a 3/5 chance that there will be a loss of £2 million. Work out the expected value of the profit or loss on the project.

3.4 Decision trees

In working out the following example, assume that all values are present values.

1 A large pharmaceutical company has developed what it considers to be a new cure for AIDS and needs to decide whether to go ahead with such a risky venture. It is estimated that it will cost £200 million to set up the plant to produce the drug; there is a **60%** chance that the project will fail, in which case the company would lose their £200 million. However, there is a **30%** chance that it will work and produce a £100 million net profit (over and above their £200 million investment) and a **10%** chance that it will yield spectacular results, with a massive £1,000 million net profit.

What decision should be made? Draw a decision tree to illustrate your answer.

2 A chain of coffee bars has just lost one of its experienced managers and will be without a full-time manager for one month. Executives of the chain have to decide whether to:

  i   Close the coffee bar for a month and make a financial loss of £2,000.

  ii   Give temporary responsibility for running the coffee bar to an existing employee with little management experience for one month. The existing employee would need to be paid an extra £2,000. The chance of their achieving a healthy profit, i.e. £8,000, is 0.2. The chance of their achieving a modest profit, i.e. £4,000, is 0.8.

iii Recruit an experienced temporary manager from an agency for one month. This would involve a payment of £8,000, made up of the manager's salary for one month plus the agency commission. The probability of this manager making a good profit of £10,000 is 0.9, while the probability of their making a modest profit of £6,000 is 0.1

Calculate the expected values of each of the three alternatives.
Which is the most desirable alternative in terms of financial criteria?

3.5 The normal distribution

1 a What is normal distribution?
   b What percentage of observations lie within **one** standard deviation of the mean, and **two** standard deviations of the mean, given normal distribution?

2 An onion-growing co-operative in the South of France grades and prices *ognions doux* (sweet onions), according to their diameter. The following table shows the price of onions with different diameters:

Price per onion (cents)	Diameter (cm)
4	Below 5
5	Above 5 and below 6
6	Above 6 and below 7
7	Above 7 and below 8
8	Above 8 and below 9
9	Above 9 and below 10
10	Above 10

Evidence shows that the mean diameter of the onions is 7.5 cm with a standard deviation of 1 cm.
   a What percentage of onions will be priced at each of the values 4, 5, 6, 7, 8, 9 and 10 cents.
   b What would be the total price of 10,000 onions?
   c What is the mean price of the onions?

3 a Three companies, A, B and C have roughly the same number of employees and the distributions of their salaries have the same modes. The distributions for A and B are approximately normal, but that of B has half the standard deviation as that of A. The distribution for C is positively skewed, but with the same range as that for A. Sketch the three distributions, using the same axis for all three.
   b If you were an ambitious 'high-flyer', which company would you prefer to work for, and why?
   c The distribution of A's salaries is shown below:

Salary range (£)	No. of employees
14,000–15,999	9
16,000–17,999	13
18,000–19,999	285
20,000–21,999	433
22,000–23,999	192
24,000–25,999	39
26,000–27,999	9
	1,000

   i Estimate the median salary.
   ii The mean salary is £21,000 and the standard deviation is £2,000. Determine whether the distribution is approximately normal. What assumptions have you made?
   iii How might the firm's employees use government–published statistics in preparing their case for a salary increase?

You may find the following information of use:

Standard deviations from mean	0–1	1–2	2–3
% of area under normal curve	34	14	2

3.6 Sampling

1 What is a representative sample?
2 What is the relationship between sampling and normal distribution?

UNIT 4
RELATIONSHIPS BETWEEN VARIABLES

Topics covered in this unit

4.1 Scatter graphs
How to use a scatter graph to plot **bi-variate associations** – i.e. the relationship (or non-relationship) between two variables plotted on an *x*- and *y*-axis.

4.2 Correlation
Defining positive and negative correlation; using the **correlation coefficient** as a measure of the correlation in a bi-variate relationship.

4.3 Regression
How to use **regression analysis** to predict the value of one variable from the value of the other variable.

4.4 Forecasting
Developing 'what if' scenarios for prediction purposes; analysing the effect of changing an independent variable on a dependent variable.

Many of the variables that are studied in business and economics are closely related. For example, when economic activity in the UK accelerates, the amount of money that people spend on foreign holidays usually increases proportionally.

Some variables directly *cause* changes in other variables – for example, when productivity increases, a firm's profitability will increase as a result.

Studying the statistical relationships between variables enables us to better explain, understand and forecast the impact of changes in specific variables.

4.1 SCATTER GRAPHS

Key Ideas

Relating two variables

Income and Consumption Expenditure by Families

Income	Consumption
£000s	*£000s*
5	8
40	32
15	20
30	25
50	40
10	12
6	10
60	45
20	20
25	24

A statistical investigation often involves the observation of two variables that are related in some way. For example, the amount spent on consumables by families is often related to their disposable income.

In order to explore the relationship between two variables, the first step is to collect data and to set them out in a table. For example, the table (*opposite*) shows data for family income and the amount of consumer spending by each family over a year.

Once the information has been collected it can be set out in what is known as a **scatter graph**. On this graph one variable is measured from the horizontal axis and the other from the vertical axis. Each intersection of the two measurements represents an observation, and each observation shows the relationship between the two variables.

When constructing a scatter graph, it is important to choose the scales carefully so that the spread of points horizontally is roughly the same as the spread of points vertically. The diagram will therefore be roughly square in appearance.

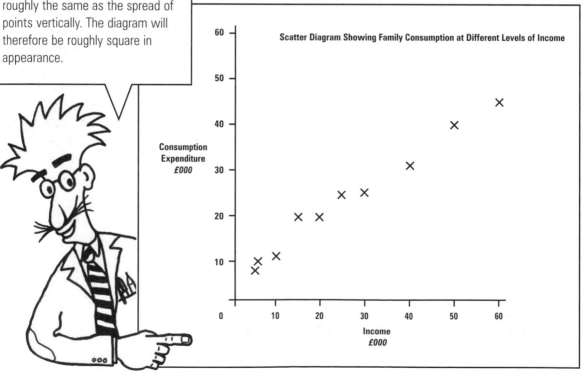

4.1 SCATTER GRAPHS

Key Ideas

Line of best fit

Where the relation between two variables is a simple or direct one, the result can often be represented by a **straight line** sloping down from left to right or vice versa.

Using a computer package such as Microsoft Excel, it is easy to set up a scatter graph and then simply to ask the program to draw the line of best fit for you.

For example, a graph showing amounts withdrawn from a bank account on one axis and the resulting bank balance on the other axis would look like this:

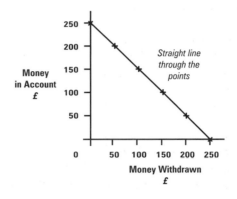

However, in the real world, relationships are rarely as exact as this. In plotting points on a graph, we often find that they are scattered on either side of a straight line drawn to show the typical relationship between the two variables.

The straight line which can be drawn through the dots to show the 'typical' relationship between the two variables is called the **line of best fit**. This is a **measure of central tendency** which seeks to iron out the variations between the scattered points.

In the diagram below, the line of best fit shows the central tendency in relating family income to family expenditure. It shows the straight line from which the scattered points deviate.

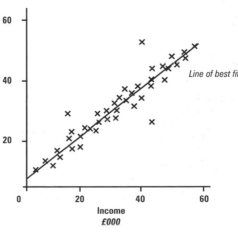

4.2 CORRELATION

Positive and negative correlation

We use the term **correlation** to describe the relationship between two variables.

Correlation refers to situations in which there is an association between the behaviour of two variables.

Doctor Proctor outlines... TYPES OF CORRELATION

- When both variables are moving consistently in the same direction – e.g. consumption increases as income increases – we say there is a **positive correlation**.

- When one variable moves consistently in the *opposite* direction to the other variable (e.g. one falls while the other rises), we say there is a **negative correlation**. (For example, as my withdrawals increase, the amount of money in my bank account falls.)

- When there is no pattern to the relationship between two variables we say that there is **no correlation**.

If points in a scatter graph cluster close to the line of best fit, they are said to show a **strong correlation**; if they are more widely scattered, the correlation is **weak**.

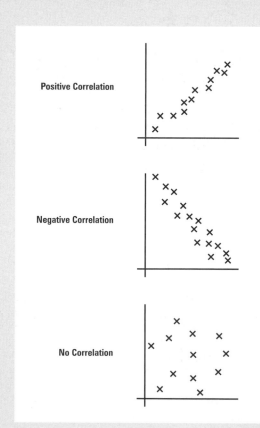

Positive Correlation

Negative Correlation

No Correlation

Strong Correlation
(tighter)

Weaker Correlation
(looser)

4.2 CORRELATION

Key Ideas 🔑

Correlation coefficient

The **correlation coefficient** is a numerical measure of the strength of a correlation. It is represented by the letter r. It measures the closeness to which a sample of paired values fit a straight line.

The most common form of correlation coefficient is the **Pearson correlation coefficient**, which measures the extent to which each value differs from the mean of its own distribution, the standard deviation of the two distributions, and the number of pairs of values.

Most computers and calculators include a function for calculating Pearson's coefficient. The formula is shown (*right*).

Doctor Proctor Calculates ...

The formula for the Pearson correlation coefficient is:

$$r = \frac{XY - n\overline{X}\overline{Y}}{\sqrt{((\sum X^2 - n\overline{X}^2)\ (\sum Y^2 - nY^{\overline{2}}))}}$$

where X and Y represent each observation of the two variables involved.

The value of **r** varies from $+1$ for perfect positive correlation to -1 for perfect negative correlation.

In both cases of perfect correlation, all the points would lie on the imaginary straight line. A zero value of r indicates that X and Y are independent of each other.

Doctor Proctor outlines... MEASURING CORRELATION

The following illustration, which is widely used in statistics, provides a good way of visualising different strengths of correlation:

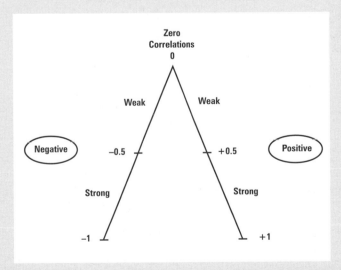

A rough guide to the strength of correlation (both negative and positive) is sometimes given as :

0.0–0.2 = very weak
0.2–0.4 = weak
0.4–0.7 = moderate
0.7–0.9 = strong
0.9–1.0 = very strong

4.2 CORRELATION

Key Ideas 🔑

Uses of correlation analysis

The purpose of **correlation analysis** is to show the association between data – for example, to show the association between the increase in sales of two separate products.

Very often decision-makers need to identify the causal relationship between one variable and another, e.g. the relationship between interest rates and consumer expenditure, income and consumption, or between advertising expenditure and sales of a product.

Doctor Proctor outlines... DEPENDENT AND INDEPENDENT VARIABLES

Where the magnitude of one variable depends on the other, we say that:

- the variable that *causes* the change in the other is the **independent variable**

- the variable that *responds* to the change in the other is the **dependent variable**.

For example, the Chancellor of the Exchequer may want to analyse the relationship between an increase in taxes on spending (*independent variable*) and consumer spending patterns (*dependent variable*).

However, we need to be careful in interpreting the relationship between variables. Often, further probing shows that the 'independent' variable is also dependent on the other variable. In some cases there may be a third variable that affects the other two.

The **dependent variable** *responds* to the change in the other variable.

Distinguish Between ...

The **independent variable** *causes* the change in the other variable.

4.3 REGRESSION

Plotting the line of best fit

We have already seen that when plotting a scatter graph it is often useful to fit some kind of curve to the data in order to give a clearer visual representation of the association.

Often the choice of curve will be a matter of experience, or of picking out what seems obvious from the diagram.

A straight line is the easiest to fit, and since it often gives a good fit to the data within a limited range, it is widely used. In some cases, a straight line drawn freehand is enough to make clear the relationship between the variables.

However, if you want to obtain the 'best fit' line, it may be sensible to calculate what is called the **least squares** or **regression line**. Regression lines are a more accurate technique used for calculating the line of best fit.

The diagram *(left)* shows the relationship between consumption and family income. You can see that there is a positive correlation between consumption and income: the higher the level of income, the more likely the family is to spend more money on consumables.

The equation that governs the line of best fit in the diagram is:

$$C = 10{,}000 + \frac{Y}{2}$$

where C (plotted on the y-axis) represents consumption, and Y (on the x-axis) represents income.

Using this regression line we can make predictions. For example:

If Income (Y) is £60,000, then Consumption (C) will be £40,000.

Family Consumption £000

$C = 10.000 + \frac{Y}{2}$

60
50
40
30
20
10

0   10   20   30   40   50   60

Family Income £000

Dr Proctor says:

'You Must Know This!'

In general, if we call the variable we wish to predict the y variable, and the variable we wish to do the predicting the x variable, the linear regression equation for 'y on x' is:

$$y = a + bx$$

Here the symbol b represents the slope of the line, and is also sometimes called the **regression coefficient**. The symbol a represents the **intercept** (i.e. the value of y where the line crosses the y-axis).

4.4 FORECASTING

Key Ideas

'What if?'

Forecasts set out to answer the question 'what if?' For example, 'If interest rates fall by 2%, what will be the effect on the average firm?'

In order to answer the question, we first need to gather research evidence. We might survey a representative sample of 100 firms and ask them what they would do if interest rates fall. Based on the results of the survey we could reach some conclusions about the behaviour of the total population.

Doctor Proctor outlines... COPING WITH UNCERTAINTY

In many cases forecasting is concerned with predicting how a change in one variable will affect another variable. How much will sales change as a result of a fall in price? How much will consumption increase as a result of a rise in income? – and so on. However, it is not an exact science: it is simply a way to cope with uncertainty.

Forecasting is based on the premise that there is an association between two variables that can be predicted from given data.

For example, it may be possible to predict that the more money a business spends on training new recruits, the higher the productivity of these employees will be. Research evidence on existing recruits may reveal the pattern shown in the graph below.

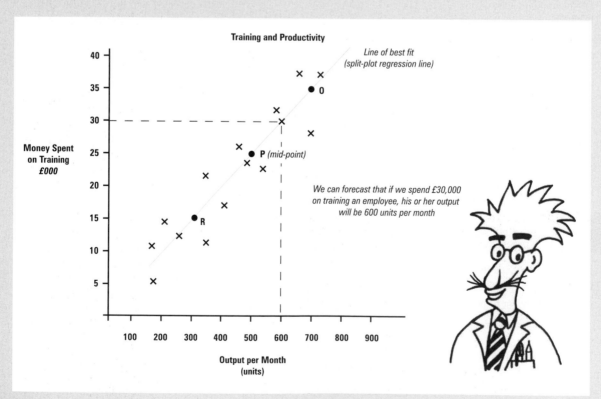

Training and Productivity

Line of best fit (split-plot regression line)

O

P *(mid-point)*

R

Money Spent on Training
£000

Output per Month
(units)

We can forecast that if we spend £30,000 on training an employee, his or her output will be 600 units per month

contd

4.4 FORECASTING

Doctor Proctor outlines... COPING WITH UNCERTAINTY (CONTD)

In statistical jargon we can say that 'productivity is the independent variable, and sales performance is dependent on it'. If this cause-and-effect relationship is strong enough, then predictions from money spent on training to productivity should be possible.

There therefore exists what we can term a **regression** of productivity upon training expenditures.

We can describe this relationship as:

Productivity = *a* + *b* (*x* Training Expenditure)

A simple way to calculate the formula for the slope of this line is to identify a point (P) that lies at the middle of the line, based on mean productivity and training expenditures. A further two points can then be plotted (**split plots**) to represent **above-average** (O) and **below-average** (U) scores.

Joining these three points together then gives us a straight line. The slope of the line thus gives us a **split-plot regression line** that can be used for prediction. For example, we can forecast that if we spend £30,000 on training an individual, then that employee will be able to produce 600 units per month.

The slope of the line (*b*) is thus:

$$b = \frac{\text{Change in Output}}{\text{Change in Training Expenditure}}$$

This approach to forecasting is very useful: it helps a business or economics forecaster to predict the possible effect of changing an independent variable.

For example, it may help the Chancellor of the Exchequer in making decisions about interest rates.

However, we always need to be cautious. Complex relationships usually involve more than two variables. In the case examined here, no two individuals will benefit from training in exactly the same way or to the same degree.

4.1 Scatter graphs and lines of best fit

1   The following data relates to the wage bill and turnover of ten Premier League football clubs in 2002. Plot the points on a scatter graph and draw a rough line of fit.

	Total Wages £000	Turnover £000
Arsenal	34,000	68,000
Aston Villa	24,000	36,000
Bolton	10,000	15,000
Fulham	12,000	18,000
Leeds	44,000	88,000
Liverpool	40,000	60,000
Manchester Utd	45,000	90,000
Newcastle	30,000	45,000
Sunderland	23,000	33,000
Tottenham	27,000	40,000

2   What is the value of drawing the line of best fit?

4.2 Correlation

1   Which of the following would you expect to have a positive and which a negative correlation?

   • The relationship between the demand for a product and its price
   • The relationship between the profits of a business and turnover
   • The relationship between the value of the Euro and sales by European companies to the USA
   • The relationship between the number of hours worked by employees in a factory and the total level of output
   • The relationship between money spent on training in a factory and the level of reported faults in production work.

2   The following table shows increases in unemployment and changes in manufacturing output in a random sample of ten industrialised countries in 2002.

Country	Increase in Unemployment %	Increase in Manufacturing Output %
Greece	–4	5
Turkey	10	–5
Lithuania	1	6
Finland	5	–10
Ireland	14	–15
Russia	20	–12
Ukraine	–5	–2
Belgium	12	–4
France	2	–5
Luxembourg	15	–8

   a   Set out the data on a scatter graph.
   b   Is there a significant correlation?
   c   Can you draw any further conclusions?

4.3 Regression

The table (below) sets out the relationship between the profits made by firms in an industry and turnover (sales values).

Profit £000	Turnover £000
15	100
30	250
100	950
20	150
25	195
10	45
35	280
18	100
40	350
70	650

1   Set out the information on a scatter graph.
2   Add a line of best fit to the diagram.
3   What equation would you use to describe the regression line?
4   How much profit would you expect firms with the following turnovers to make?

   • £300,000
   • £70,000
   • £600,000

4.4 Forecasting

1   Give examples of **three** relationships for which a business might want to provide forecasts.
2   How useful would **one** of these forecasts be in giving the business greater levels of control?
3   Why might forecasts not be 100% accurate?
4   If the turnover of a retailer is always twice as much as the cost of sales, what equation would you use to show the slope of the line of best fit?

What is the probability that the next item sold will bring in twice as much revenue as the cost of sale?

5   The chart below shows global market share in $ billions between 1982 and 2002. Is it possible to use this evidence to make forecasts of future music sales during the period 2003–2010?

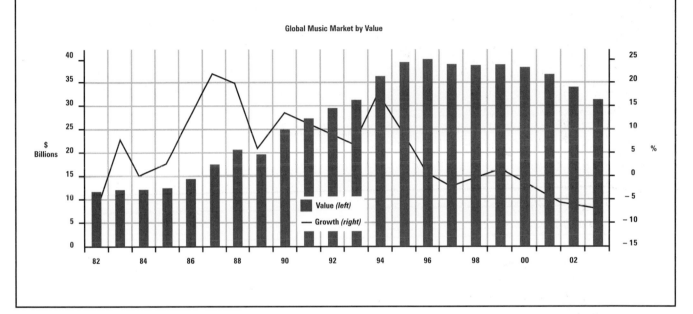

Global Music Market by Value

UNIT 5
TIME SERIES

Topics covered in this unit

5.1 Underlying trends
Analysing trends which become apparent over time and indicate longer-term general changes – for example, fluctuations in output figures.

5.2 Seasonal variations
Variations which take place seasonally (for example, demand for certain types of food such as ice cream), which cause figures to exceed or fall below their trend values.

5.3 Seasonal adjustment
How to make allowances for the effect of the seasons using an averaging process.

5.4 Forecasting future figures
How to extrapolate trends based on past and present evidence in order to forecast the future.

5.5 Cyclical and random variations
Cyclical variations are long-term variations which are difficult to interpret statistically. Random movements are assumed to have a zero effect on longer-term trends.

Time series data is made up of a sequence of values (such as quarterly output levels) which vary with time.

From the business point of view, this data is very important: it not only makes it possible to chart past performance; it can also help decision-makers to make predictions about the future. Time series are also important in analysing economic trends and in forecasting future trends.

This unit examines how time series data can be affected by four main types of external factors:

- Trends (**T**)
- Seasonal variations (**S**)
- Cyclical variations (**C**)
- Irregular or random variations (**R**).

5.1 UNDERLYING TRENDS

A **trend (T)** is a general direction or tendency whereby a variable appears to rise, fall or fluctuate over a period of time.

If you study the graph (*below*) you can see that from March to November 2002 there was a general trend for the average share price of the top 100 European companies to fall in value. There were some upturns and sharp downturns, but the overall trend was a steady decline in value.

FTSE Eurotop 100: Share Values

Doctor Proctor outlines... SEASONAL VARIATION

Business data which fluctuates over time, such as sales or production figures, is generally built up from a number of components. One of the most important factors is the underlying trend, but a **seasonal component (S)** often has to be taken into account as well.

For example, in retailing you would expect sales to be higher in the pre-Christmas period than in July and August, when many people are on holiday.

Seasonal variations can also be applied to much shorter periods. Supermarket takings are often much higher on Fridays and at the weekend than during the rest of the week.

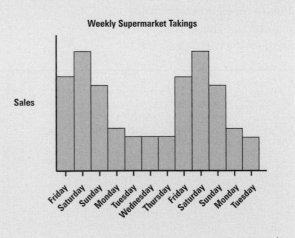

Weekly Supermarket Takings

contd

5.1 UNDERLYING TRENDS

Doctor Proctor outlines... SEASONAL VARIATION (CONTD)

In the illustration *(below)* you can see two elements causing fluctuations in sales figures. One is an **underlying trend** towards increasing sales as the company in question steadily raises its public profile and market share. This underlying trend is represented by the dotted rising line.

In addition you can see a **seasonal trend** for increased sales in December (pre-Christmas) and January (due to the January sales).

On the graph, Point X is higher than Point Y for two reasons:

- because of the upward trend over time
- because X is in December while Y is in the summer period.

High street retailers often experience seasonal fluctuations in demand. For example, trade is nearly always brisk during the 'Christmas rush' and January sales.

5.1 UNDERLYING TRENDS

Key Ideas 🔑

Eliminating random fluctuations

In addition to seasonal fluctuations, there can be a number of **random fluctuations** which affect data. To identify a trend we need to remove both the seasonal and these random influences.

We can therefore say:

Trend = Data – Seasonal Effect – Random Error

Or:

$$T = (T+S+R) - S - R$$

where T is the trend, S the seasonal effect and R the random error.

Doctor Proctor outlines... SEMI-AVERAGES

A manufacturing firm employs extra workers to work overtime when orders pick up. During the last 11 months the average number of workers working overtime has been:

4 6 8 7 10 9 10 14 15 16 18

Clearly there is a trend here to employ more workers to work overtime. But how can the trend be measured?

One approach is to use **semi-averages**.

The values of the variable (in this case, the number of workers) are divided into two halves. If there is an odd number, the middle value can either be left out or included in both halves. We then work out the mean of each of the two sections of data as follows:

$$\frac{4 + 6 + 8 + 7 + 10}{5} = 7.0$$

$$\frac{10 + 14 + 15 + 16 + 18}{5} = 14.6$$

These two mean values can be plotted on a graph at the time intervals corresponding to the mid-points of the two halves. These points can then be joined together to give a straight line which represents the trend.

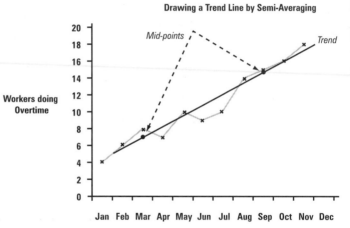

Drawing a Trend Line by Semi-Averaging

Using semi-averages is a simple way to plot a trend.

5.1 UNDERLYING TRENDS

Key Ideas

Moving averages

Where a trend is **linear** (i.e. can be represented by a straight line) the semi-average method is appropriate.

However, if the trend is not linear, it is better to use a **moving average**.

When creating a moving average, it is always important to think carefully about the basis on which your averages are calculated.

Using the four seasons of the year may be appropriate when there are clear quarterly patterns. But in the case of a supermarket, where there are distinct *daily* sales patterns, averages would need to be worked out over a seven-day period to allow for the effect of higher sales on Fridays and at weekends.

Example

A firm supplying oil for central heating faces high demand in autumn and winter and lower demand in spring and summer.

In order to find out the long-term trend in demand for its product, the business needs to eliminate the seasonal variation from its demand picture. It does this by creating a moving average which smooths out demand for the four seasons of the year.

The first step is to calculate the average for the first group of four seasons. Then an average is taken for a second group of four seasons, starting with spring, and so on. The demand patterns and resulting moving averages are shown in the diagram (*right*).

You can see from this that in each of the seasons in Year 2, values have risen by 10 from the previous year. Having removed the seasonal variation, it is therefore possible to identify a steady upward movement.

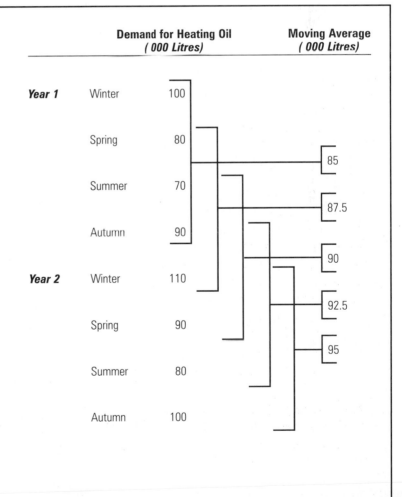

		Demand for Heating Oil (*000 Litres*)	Moving Average (*000 Litres*)
Year 1	Winter	100	
	Spring	80	
			85
	Summer	70	
			87.5
	Autumn	90	
			90
Year 2	Winter	110	
			92.5
	Spring	90	
			95
	Summer	80	
	Autumn	100	

5.2 SEASONAL VARIATIONS

Key Ideas 🔑

Additive and multiplicative methods

There are many cases where seasonal variations can distort the trend pattern of a variable over time. Obvious examples are the demand for holiday bookings or for consumer goods such as ice cream or cooling fans.

The additive approach

$$Y_n = T_n + S_n$$ (where **n** is the quarter under consideration)

Appropriate when deviations from the trend (caused by seasonal variations) are of a similar *absolute* magnitude from one peak or trough to the other peaks and troughs in a series.

Distinguish Between ...

The multiplicative approach

$$Y_n = T_n \times S_n$$

Appropriate when deviations from the trend (caused by seasonal variations) are of a similar *percentage* from one peak or trough to another.

In the **additive approach** to time series, it is assumed that the change in a variable **Y** (say, sales revenue) is the result of adding together the trend (**T**) and the seasonal variation (**S**). In other words:

$$Y_n = T_n + S_n$$ (where **n** is the quarter under consideration)

The additive method is appropriate when deviations from the trend (caused by seasonal variations) are of a similar *absolute* magnitude from one peak or trough to the other peaks and troughs in a series – for example, when every summer, sales increase by 100, and when every winter, they fall by 100, compared with the trend.

However, where deviations from the trend are of a similar *percentage* from one peak or trough to another – e.g. a 10% increase in summer and a 10% fall in winter – we may need to use a **multiplicative approach** instead.

The multiplicative approach is :

$$Y_n = T_n \times S_n$$

5.2 SEASONAL VARIATIONS

Key Ideas 🔑

Positive and negative variation

The graph shows sales of a brand label T-shirt in different quarters of the year (winter, spring, summer and autumn). You can see that demand is highest in summer, and lowest in winter. However, the trend line shows a steady increase in demand over time.

In any one quarter, the trend value of sales may be up or down compared to the previous quarter. However, the seasonal variation may be either **positive** or **negative**, depending on whether actual sales figures are above or below the trend value in that quarter.

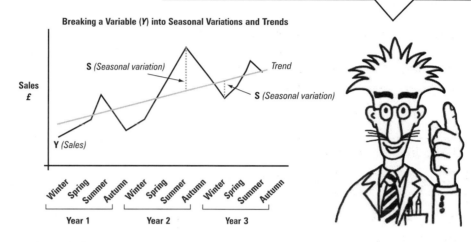

Breaking a Variable (*Y*) into Seasonal Variations and Trends

---

Doctor Proctor outlines... SIMPLE UNDERLYING TRENDS

When examining the effect of seasonal variations we need to identify the relevant seasons. For example, if our observations relate to the four seasons of the year, a **four-point moving average** would be appropriate, e.g. for winter, spring, summer and autumn.

Let's take a simple example in which an ice cream seller is faced by falling demand for ice cream in the winter and rising demand in the summer. Each year he sells 5,200 ice cream cones: 2,000 in the summer, 1,200 in spring and autumn, and 800 in the winter (*see table*).

The mean value for any sequence of four seasons will be 1,300. The *trend* therefore is the same as the *mean*: it is neither rising nor falling, but steady at 1,300. The trend is therefore a straight line.

Seasonal variations lie above the mean (summer, 2,000) or below the mean (spring, 1,200; autumn, 1,200; winter, 800).

In the real world it is more common for the trend to move around. Sometimes the long-term trend will be rising, sometimes falling. You would expect to see a rising trend in demand for a new product that outperforms its rivals.

	Year 1				Year 2			
Spring	*Summer*	*Autumn*	*Winter*	*Spring*	*Summer*	*Autumn*	*Winter*
1,200	2,000	1,200	800	1,200	2,000	1,200	800

If we take a moving average over any four quarters it will be:

$$\frac{1,200 + 2,000 + 1,200 + 800}{4} = 1,300$$

5.2 SEASONAL VARIATIONS

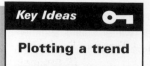

Key Ideas

Plotting a trend

Plotting a trend simply involves finding moving averages which are relevant to the sequence of quantities that you are studying.

If you are studying figures where there are four seasons, then you take moving averages covering every four sets of figures.

Example

The sales of a product are as follows:

Year	Quarter	Sales
1	1	100
	2	126
	3	124
	4	96
2	1	88
	2	106
	3	96
	4	60
3	1	44
	2	54
	3	52
	4	24

We can now set out these figures as a moving average. For example, for the first four months shown the moving average is:

$$\frac{100 + 126 + 124 + 96}{4} = 111.5$$

We need to plot this point halfway down the set of values, i.e. between the second and third values. However, as this does not correspond to any of the *actual* values, it is helpful to create a centred value – halfway between the moving average of the first four values and the moving average of the second, third, fourth, and fifth values, as shown in the table (*below*).

Year	Quarter	Sales	Moving	Centre of Moving Averages (Trend)
1	1	100		
	2	126		
			111.5	
	3	124		110
			108.5	
	4	96		106
			103.5	
2	1	88		100
			96.5	
	2	106		92
			87.5	
	3	96		82
			76.5	
	4	60		70
			63.5	
	1	44		58
			52.5	
	2	54		49
			43.5	
	3	52		
	4	24		

A moving average compares average values taken over relevant periods of time. For example, a seasonal moving average consists of an average for winter, spring, summer and autumn, followed by an average for spring, summer, autumn and winter – and so on.

5.2 SEASONAL VARIATIONS

Example (contd)

You can see from this that sales are highest in the second quarters of each year and lowest in the fourth quarters. However, to get rid of the seasonal variation and to smooth out the trend, we simply create a moving average and centre up the values of the moving average.

We can now set out the table again to illustrate seasonal variation. To do this we look at the **deviation** (difference) between the actual sales figures, and the **trend** (centred moving average).

The deviation now shows the seasonal variation.

Year	Quarter	Actual Sales	Trend	Deviance (Seasonal Variation)
1	1	100		
	2	126		
	3	124	110	+14
	4	96	106	−10
2	1	88	100	−12
	2	106	92	+14
	3	96	82	+14
	4	60	70	−10
3	1	44	58	−14
	2	54	49	+5

Seasonal variations are shown by the deviation between actual figures and smoothed trend figures.

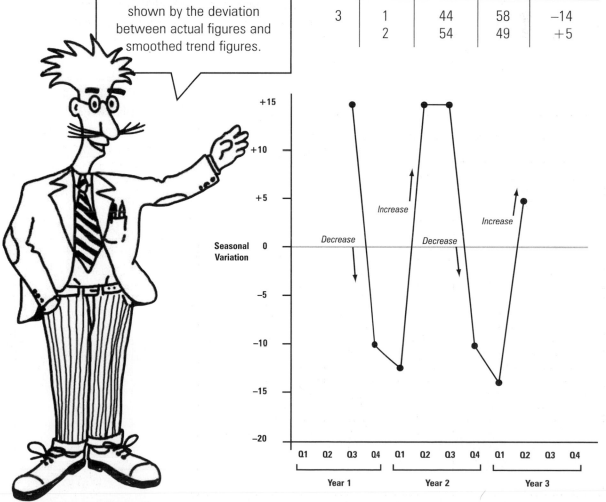

5.3 Seasonally Adjusted Figures

As we have seen, in examining sets of data it is very important to take account of seasonal effects.

The unemployment figures for the summer of 2002 may look particularly bad until we notice, for example, that they are lower than at the same time the previous year – just as the rise in unemployment in the summer of 93 was less than in the summer of 92 and 91. It is quite usual for unemployment to rise when school-leavers and university students swell the numbers of those looking for work.

Statisticians need to be aware of – and adjust for – these seasonal patterns.

Doctor Proctor outlines... 'Smoothed' Data

Most economic data – and a lot of business data such as sales and employment figures – is produced monthly and quarterly. This data is often obscured by regular seasonal events such as changes in the weather, opening and closing of schools or major holidays.

Because of the British climate, for example, employment in the building industry typically slows down in winter and picks up again in spring. Given that this increase is seasonal, how do we know if an increase in building employment is higher or lower than normal?

Removing or neutralising regularly occurring increases or decreases during a given month results in 'smoothed' data that give us a better perspective from which to see the true size of the over-the-month change.

For example, the table (*below*) shows the difference between unadjusted and seasonally adjusted data for the building industry in a specific region of the UK in 2002.

Example

Numbers Employed in the Building Industry in Region X, Jan 2002 – Dec 2002 (000s)

Month End	Seasonally Adjusted Unemployment	Unadjusted
Jan	99.1	82
Feb	100.2	81
Mar	99.9	82.5
Apr	101.4	93.2
May	101.1	103.7
Jun	100.9	111.7
Jul	101	115.1
Aug	101.3	116.4
Sep	101.9	114.2
Oct	103.3	113.3
Nov	104.5	107.9
Dec	106.5	100

5.3 SEASONALLY ADJUSTED FIGURES

Key Ideas

Month-by-month adjustment

Because seasonal events affecting employment such as weather and major holidays often follow a more or less regular pattern each year, their influence on statistical trends can be eliminated by adjusting the statistics from month to month.

These adjustments make it easier to observe the long-term trend. However, it must always be remembered that seasonal adjustment is an approximation and that initial adjustment must be based on past experience.

Doctor Proctor outlines... METHODS OF SEASONAL ADJUSTMENT

Seasonal adjustment usually involves using one of the following techniques:

- Several years of historical experience – often as much as seven to ten years' worth of data – are used to develop a **seasonal adjustment factor**.

- The most recent year's seasonal factors are 'added in' every year, replacing the seasonal factors used in previous years. For example, if seasonal adjustment is being carried out over a seven-year period, this month's data will replace the data from the same month seven years ago.

- Extreme values for a particular month are excluded as aberrations or are given less weight. For example, in measuring employment figures, we may want to exclude a figure for a month in which there is widespread industrial action.

- Seasonal patterns of recent years are given more weight than seasonal patterns from earlier years.

Don't forget that seasonal adjustment is an approximation and that adjustment should always be based on past experience.

5.4 FORECASTING FUTURE FIGURES

Key Ideas 🔑

Allowing for the unexpected

Predictions are statements which are *probable* or *credible* in that they attempt to take the 'don't know' out of economic and business decision-making and replace it with 'I think' or 'I predict'.

The best that forecasting provides us with is a view of the future on which it may be possible to construct plans.

Forecasting involves identifying past trends, seasonal variations and other predictable patterns and using these patterns to make **predictions** about the future.

Of course, it is more difficult to forecast the future than to chart the past, because the future can never be certain.

Doctor Proctor outlines... TRENDS, SEASONAL VARIATION AND RANDOM VARIATION

In business forecasting, it is possible to establish expected values based on careful analysis of trends and other fluctuations.

One formula for forecasting business data such as sales, profits and production figures is:

Forecasted figures $= T + S + R$

where T is the expected trend, S is the expected seasonal variation and R is a random variation.

1 Establishing the trend

In establishing a trend we need to examine the most recent patterns and any causal explanations that may underlie them.

The process of extending a trend outside the time-range of the original data is called **extrapolation**.

Estimates of future figures can be based on extending the trend and seasonal variations from previous periods.

2 Establishing the seasonal variation

In establishing the seasonal variation, we need to examine the most recent seasonal variations, and examine whether these are likely to change over time.

3 Estimating the random variation

The random variation is the most difficult to estimate. One method is to calculate random changes on the basis of how significant random factors have been in the past. Sometimes it makes sense to ignore random factors or discount negative random factors altogether. If we are feeling pessimistic, however, we may want to make a generous allowance for negative random factors.

Forecasts may be based on moving averages, i.e. trends from the past, or we may want to build in negative factors by allowing for, say, a likely fall in sales figures of 2% a month.

5.5 CYCLICAL AND RANDOM VARIATIONS

Key Ideas

The business cycle

Cyclical and random factors are the most difficult to predict. In this book, therefore, we have simply stated that these factors exist and described them briefly. At this introductory level, it is more important to understand basic statistical principles such as how to calculate a trend or the seasonal components of a time series.

However, many aspects of business and economic activity *are* affected by **cyclical variations**, including demand for products, employment, and national patterns of consumer expenditure. It is difficult to predict these in a reliable way because the cycle is liable to vary in length and intensity and is affected by many external factors *(see below)*.

Again, **random movements** are almost impossible to predict, and are assumed to have a zero effect on the trend in the longer term.

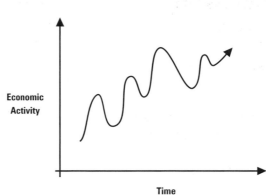

The Business Cycle

Economic Activity

Time

The diagram *(above)* shows that economic activity can fluctuate dramatically.

Although the alternation of boom and slump forms a predictable pattern, the short-term effect is hard to predict because the different phases of the cycle are liable to vary in length and intensity.

Doctor Proctor outlines... ALLOWING FOR BOOM AND SLUMP

The business cycle is made up of periods of **boom** and **slump** in economic activity. These periods may last two or three years or even longer.

In forecasting the effects of these booms and slumps, statisticians are wise to take into account the 'best estimates' of experts and build these into their forecasts.

However, substantial changes in economic activity remain very difficult to predict. For example, few could have anticipated the September 11th atrocity and the global downturn in economic activity which followed it.

5.1 Underlying trends

1 Sales figures at Mucky Breweries this month have risen by 20% compared to the previous month. The sales manager argues that this shows a rising trend. But is this true? What further evidence would be needed in order to prove the case?

2 The following data shows the direct cost (in pence) of producing a bar of chocolate between 1992 and 2002.

Year	Direct Cost
1992	2.0
1993	2.5
1994	2.5
1995	3.0
1996	3.0
1997	3.5
1998	4.5
1999	6.0
2000	6.5
2001	7.5
2002	8.0

Work out the five-year moving average and calculate the mean annual increase in cost between 1994 and 1996, between 1998 and 2000, and between 1994 and 2000.

5.2 Seasonal variation

1 When calculating seasonal variations, how do you decide on what constitutes a season?

2 A clothing company has calculated that demand for one of its items nationally is increasing by 10,000 per quarter. At the same time demand for this item is subject to a seasonal variation, increasing in the summer and winter months. Draw a graph to illustrate the seasonal variation using the figures below.

	Demand (000s)	
Year 1	Spring	20
	Summer	40
	Autumn	40
	Winter	60
Year 2	Spring	60
	Summer	80
	Autumn	80
	Winter	100

5.3 Seasonally adjusted figures

The table (*below*) shows unemployment figures in a London borough in 2002.

	Numbers unemployed
April	5,000
May	4,900
June	4,800
July	6,000
August	6,400
September	6,000
October	5,000

1 Why might these figures 'exaggerate' the unemployment situation?

2 How could these figures be adjusted to give a more accurate picture of trends in unemployment?

5.4 Forecasting future figures

1 Demand for a product is currently (May) 50,000. The trend has been for demand to rise by 1,000 a month for the last two years. In December and January there is always a further increase in demand of 2,000 in December, and of 1,000 in January. Set out a table giving a forecast of demand between May this year and May next year.

2 What is the difference between an additive and a multiplicative approach to forecasting?

5.5 Cyclical and random variations

1 What is meant by the **business cycle**? Why is it more difficult to forecast cyclical than seasonal patterns?

2 How do we usually account for random factors in setting out a trend?

Topics covered in this unit

6.1 Simple index numbers

Using simple indices rather than actual values in order to show changes in a variable over time.

6.2 Complex index numbers

Using more complex index numbers which use weighting to reflect the relative importance of different items.

6.3 Important published indices

How two of the most widely quoted indices – for share values and retail prices – are calculated and used.

In the financial press you will often see references to **indices** – for example, in coverage of share price movements, exchange rate fluctuation and movement in other economic variables.

Business specialists also frequently use index numbers to present and summarise financial results such as sales and profit figures.

In this unit we look at how indexes are constructed, and give examples of how they are typically used in the business world.

6.1 SIMPLE INDEX NUMBERS

Key Ideas 🔑

The advantages of indices

Index numbers are a convenient way to make comparisons of changes in quantities over time.

Instead of measuring changes in a variable in terms of units such as '£s of sales' or 'tonnes of output', values are expressed in abstract numbers or ratios, rather like percentages.

Index numbers can also be set out in a time series to illustrate how sales, production levels, or prices have changed over time. This conveys an overall picture much more clearly and directly than a chart or table based on the original units of measurement.

Dr Proctor says:
'You Must Know This!'

To calculate an index value, we first need to establish a **base position**, or benchmark, from which changes can be measured. This is given an index number of 100.

The index values are calculated by dividing the new value by the base value, multiplying by $\frac{100}{1}$ and adding 100.

If the base position is set at 50, an actual value of 60 is therefore given an index number of 120.

Doctor Proctor outlines... CONSTRUCTING A SIMPLE INDEX

In constructing an index, we need to start with a base value of 100 against which comparisons can be made.

For example, to monitor sales figures we might establish a starting point at the year 2000, when sales were £50 million. Using the method outlined *(above)*, sales figures for the following years could then be given an index value as shown in this table:

Year	Actual Sales *(£m)*	Index Number
2000	50	100
2001	60	120
2002	70	140
2003	100	200

One of the main advantages of setting out figures in this way is that it helps us to make comparisons between the relative rate of change of different variables.

For example, even though the exports and imports of a country such as the UK are vastly different (exports include whisky and exams, while imports range from wine to televisions), we can still make comparisons of changes in export prices compared with import prices.

Using index numbers it is possible to measure the **terms of trade** – i.e. the relationship between export prices and import prices – as follows:

The terms of trade =

$$\frac{\text{Index of export prices}}{\text{Index of import prices}} \times \frac{100}{1}$$

6.1 SIMPLE INDEX NUMBERS

Key Ideas 🔑

Calculating percentage changes

Once an index number has been assigned to a set of variables, it is possible to calculate the percentage change from one index number to another.

This is done using the formula:

$$\% \text{ Change} = \frac{\text{New Index} - \text{Old Index}}{\text{Old Index}} \times 100$$

Example

During the past year index values for industrial production have gone up from 120 to 140.

This can be expressed as a percentage change as follows:

$$\% \text{ Change} = \frac{140 - 120}{120} \times 100$$

$$= \textbf{16.66\%}$$

Doctor Proctor outlines... A SIMPLE PRICE INDEX

Suppose that we want to compare changes in the price of three items that feature heavily in a typical family budget: a standard loaf of bread, six eggs, and half a litre of milk. The table shows the actual price of these items in pence over a 3-year period:

Price (pence)

	2001	2002	2003
Bread	30	40	60
Eggs	50	60	80
Milk	100	110	120

To create a simple price index we need to calculate the ratio of the new prices to the base year for each item. The ratio of the new price to the original (base year) price we call the **price relative**.

A formula for a simple price index is:

$$\text{Index} = \frac{P_n}{P_0} \times 100$$

where P_n represents the price in a particular year (Year n), and P_0 represents the base year price (Year 0).

We can set this out in the new table (*below*).

You can see that during the period covered in the table, bread has gone up in price by 50%, eggs by 60% and milk by 20%.

Year	Bread Price	Bread Pn/Po	Bread Simple Price Index	Eggs Price	Eggs Pn/Po	Eggs Simple Price Index	Milk Price	Milk Pn/Po	Milk Simple Price Index
2001	30	1.00	100	50	1.00	100	100	1.00	100
2002	40	1.33	133	60	1.20	120	110	1.10	110
2003	60	1.50	150	80	1.60	160	120	1.20	120

6.1 SIMPLE INDEX NUMBERS

Key Ideas 🔑

Calculating an aggregate price index

In the example shown on page 85, we looked at the increase in price of three separate items: a standard loaf of bread, six eggs, and half a litre of milk.

But it is also possible to work out a simple price index for the three items combined. This is known as a simple **aggregate** price index.

The sum of the three prices is represented by the symbol $\sum$. The simple aggregate price index is therefore =

$$\frac{\sum P_n}{\sum P_0} \times 100$$

The new calculations are set out in the table (*below*).

Price in Given Year (Pence)	Bread	Eggs	Milk	$\sum P_n / \sum P_0$	Simple Aggregate Price Index
P(2001)	30	50	100	180/180=1.0	100
P(2002)	40	60	110	210/180=1.16	116
P(2003)	60	80	120	260/180=1.44	144

The problem with this approach is that it ignores the fact that the typical family will spend more in their weekly shopping on some of these items than others – e.g. more on bread than eggs.

As a result, changes in the price of eggs are being given more significance than they warrant, while changes in the price of bread are being given *less* significance than they warrant.

To avoid this distortion, it is necessary to introduce a form of **weighting** that takes into account the relative importance of different items in the aggregate index. The principles of weighting are explained on page 87.

6.2 COMPLEX INDEX NUMBERS

'Weighting' an index

To show the importance of weighting let us look at a practical example.

In order to reflect the relative importance of different items in an aggregate or composite index, it is necessary to include an element of **weighting**.

Example

A company manufacturing chairs has two basic inputs: materials and labour.

In the table (*below*) you can see that between 2002 and 2003, the amount spent by the firm on materials and labour has increased.

	2002 (1) (£)	2003 (2) (£)	Index Number 2/1 x 100	Expenditure Weight (£s)
Labour (unit cost)	50	55	110	150
Material (unit cost)	200	250	125	200

However, what the table does not show is the **relative magnitude** of the increases – i.e. which of the price increases has affected the firm most significantly. Labour unit costs have gone up from 50 to 55 and material unit costs from 200 to 250, but what impact has this had on the firm?

In order to answer this, we need to calculate what is known as the **price relative** of both labour and materials.

The unit cost of labour has risen from 50 to 55. Taking 2002 as the base year, we can calculate the price relative of labour from 2002 to 2003 by expressing it in terms of an index of unit labour costs:

$$\frac{55}{50} \times 100 = 110$$

We can do the same for materials, giving us a price relative index of 125. Immediately you can see from the index that materials have risen 2.5 times as much as labour.

We can also find the **combined** effect of the rise in price of labour and materials. This can be done using a simple averaging technique:

$$\frac{110 + 125}{2} = 116.5$$

However, this would only be a useful figure if the amount spent by the company on materials were the same as the amount spent on labour – which is not the case. From the table (by reading the expenditure weight) you can see that the firm buys 3 units of labour for £150, and 1 unit of material for £200. The simple averaging technique used above has therefore underestimated the importance of materials.

contd on page 88

6.2 COMPLEX INDEX NUMBERS

Example (contd from page 87)

More weight needs to be given to materials because materials make up a bigger proportion of total costs than wages.

The true percentage increase in total cost is obtained from the **weighted** average of the individual indices, as follows:

$$\frac{(£110 \times 150) + (£125 \times 200)}{(150 + 200)} = 118.5$$

This shows that the total cost increased by **118.5** rather than **116.5**.

This can be confirmed as follows. If 3 units of labour and 1 unit of materials are required to produce a chair, then in 2002 the total cost of producing a chair was

$$50 \times 3 + 200 \times 1 = £350$$

In 2003 the total cost was:

$$55 \times 3 + 250 \times 1 = £415$$

So the total cost in 2003 as a percentage of 2002 is:

$$\frac{415}{350} = \mathbf{118.5}$$

When the weights used in an index are assumed to remain the same as in the base year, we refer to a **base-weighted index**. This type of index may be used when – for example – a firm continues to use the same ratio of materials to labour.

When weights need to be altered over time, we refer to a **weighting-adjusted index**.

For example, in a price index based on family expenditure, it is highly likely that spending patterns will change over time, with families perhaps spending more on leisure and recreation and less on food. In this case, the index needs to be based on what families buy *now*, and these costs compared with the cost of the same basket of items in the base year.

A **base-weighted index** is one in which the weighting used remains the same over time.

Distinguish Between ...

A **weighting-adjusted index** or **current weighted index** is one in which weights are adjusted to reflect changes over time.

6.3 IMPORTANT PUBLISHED INDICES

Key Ideas	🔑

Share price indices

Share price indices are widely quoted in the business and financial press. Some of the most important are:

- The **FTSE ('Footsie') 100 Index** gives hourly market movements for the most significant 100 companies listed on the London Stock Exchange. It is used as an indicator of the health of the economy. This index started with a base of 1,000.

- The **FTSE 250 Index** is made up of the next range of companies, just below the size that would qualify them for the FTSE 100. The **FTSE 350** combines the 100 and 250 index. There are separate indices for smaller companies.

- The **FTSE TechMark** index tracks the progress of technology-related shares.

- The **FTSE All Share index** covers all listed companies.

- Within the European Union, the most important index is the **Euro Top 200**, which covers top European companies.

- The **Dow Jones Industrial Average** covers the New York Stock Exchange, while the **Nikkei 225** covers the Tokyo market.

Doctor Proctor outlines... CALCULATING A SHARE PRICE INDEX

Share price indices compare share prices over time with share prices at a particular base date.

For example, take the following three shares in the leisure industry:

Share	Price in 2002	Price in 2003
Premier Football Club	200	300
Super Leisure Fitness	100	180
New Hotels	60	90

To work out the indexed increase, we use 2002 as the base year and compare 2003 prices with this.

	Base (2002)	New Figure (2003)
Premier Football Club	100	150
Super Leisure Fitness	100	180
New Hotels	100	150

To work out the average increase we add together the three figures for 2003 and divide by the number of companies, i.e.

$(150 + 180 + 150) = 480$

Number of companies = 3

Average increase $= \dfrac{480}{3} = 160$

Thus we see an average increase in this share index of 60%.

Movements in an index are usually measured in **points**.

For example, if the FTSE 100 drops from 5,000 to 4,900, it is said to have fallen 100 points.

Share price indices tend to fluctuate with business confidence.

6.3 IMPORTANT PUBLISHED INDICES

The Retail Price Index (RPI)

The **Retail Price Index (RPI)** gives a measure of the percentage change in the general level of prices of a typical 'basket' of goods for household consumption in the UK.

Doctor Proctor outlines... COMPILING THE RETAIL PRICE INDEX

In the UK the retail prices of items such as leisure expenditure, transport and food are collected each month by government employees who visit a range of randomly selected retail outlets in different parts of the country.

The index is weighted according to the relative amount of expenditure that goes on the different items that make up the index.

The index is regularly updated to introduce new items and take out items which are no longer widely consumed.

The price ratio for individual goods and services is weighted by the average household expenditure on that good.

Household expenditure

Data on spending is also collected each year by the **Family Expenditure Survey** from a random sample of about 11,000 households from across the country.

Each member of the household keeps a record of purchases over a two-week period, and the head of the household is required to record more significant family expenditure, e.g. mortgage repayments.

Recently, leg-waxing has entered the RPI and one of my cat's favourite items, tinned salmon, has been taken out!

6.1 Simple index numbers

1 Provide a formula for a simple index.
2 The prices of four items bought by a typical household changed in the following way between 2000 and 2004:

Price (Pence)

Item	2000	2001	2002	2003	2004
Bread	80	85	90	95	100
Cheese	100	120	130	140	160
Eggs (6)	60	65	70	75	80
Wine	200	210	220	230	240
Cigarettes	400	410	440	450	500

  a Set out indices to show how the prices of these goods changed in the period shown.
  b Set out a composite index to show average price changes of the four items in each of the years, using the year 2000 as your base year.

6.2 Complex index numbers

1 Why is it important to weight items when constructing a complex index?
2 Why might it not be appropriate to weight items in the base year?
3 The table (*below*) shows prices and quantities of a basket of consumer goods over a two-year period.

Goods	2000 (Pence)		2002 (Pence)	
	P^1	Q^1	P^2	Q^2
Beans	30	200	45	200
Milk	50	500	60	500
Eggs	20	300	30	300

Calculate a price index for this basket, using the year 2000 as your base year.
4 What would be the impact of the changed pattern of consumer spending from Year 1 to Year 2?

5 How could the price index be recalculated to take account of changes in the pattern of spending in Year 2?
6 In a given year, the mean salary of the five members of an Internet-based bookselling company was £47,500. A total of 25,000 books were bought at a mean price of £4.25; services cost £60,000.
  a Determine the **weighted aggregate index** to show how the total expenditure in the given year compared with what it would have been in a base year when the mean salary was £40,000, the mean cost of a book was £3 and services cost £50,000.
  b Is the **simple aggregate index** for the given year appreciably different from the weighted aggregate index?

6.3 Important published indexes

1 What is the FTSE All Share Index?
2 What is the FTSE 100?
3 Which of the two share price indices is likely to experience the widest fluctuations?
4 On Wednesday 16 October 2002, national newspapers reported that UK share prices were soaring, after better-than-expected figures from some of America's largest corporations, including General Motors, Bank of America and Citigroup, had sent US shares into orbit. The FTSE 100 followed suit, ending 198.7 points higher at 4,236.5, while the FTSE 250 gained 125 points.
  a What is the difference between the FTSE 100 and the FTSE 250?
  b Why do you think that the FTSE 100 did better than the FTSE 250?
  c Why are changes in the FTSE measured in points?

Answers

Unit 1: Organising, presenting and interpreting statistical data

1.2 Organising, presenting, interpreting

1 Descriptive
2 Inferential
3 Inferential
4 Descriptive

1.3 Tables

Amount spent in supermarket by 80 shoppers

Amount spent (£)	0–20	21–50	51–100	101–200	Over 201
Number of shoppers	18	17	18	13	14

1.4 Pictograms

The pictogram shows that Vin Extraordinaire was most successful in selling wine to England – 8,000 bottles, followed by Scotland with 5,500 bottles and finally Wales with 3,500 bottles.

1.5 Bar charts

From studying the bar chart it is obvious that there is a clear difference in the time spent in school in groups of developed countries such as the USA, Canada, Japan, Germany, France and the UK, compared with a second group of developing countries such as Egypt, Algeria and South Africa.

The height of the bars reveals that children in western countries spend twice as much time in school as children in Egypt, and Algeria, while Senegal lags well behind even other developing African countries.

The bar chart is graphic, simple, and makes differences stand out. It is a good way of presenting information in newspapers, reports and visual presentations.

1.6 Pie charts

1

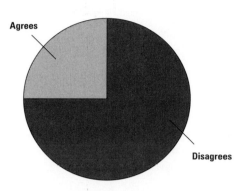

2 Plotting the data in Microsoft Excel should provide clear visual results.
3 A bar chart makes it easy to compare the different heights of the bars.

It is not always easy to make comparisons on a pie chart – for example, the segments for Digital Terrestrial (non-pay) and for Digital Cable will be hard to distinguish on a pie chart because they are similar in size (17% and 18%, respectively). It is however often easier to label pie charts than bar charts.

1.7 Line graphs

1 Shares were highest, at about 160p, in March 2002.
2 The share price has gone up 90p, from about 70p to 160p.
3 A line graph is very easy for the eye to follow, with time shown on the horizontal axis and changes in the variable shown on the vertical axis. Joining together points provides a form of averaging process, enabling changes over time to be quickly understood.

1.8 Scatter graphs

The scatter graph illustrated on page 94 seems to indicate that the fewer goals a team concedes, the more likely it is to win matches. The straight line indicates a rough line of best fit between the points.

Arsenal diverge most from the line of best fit largely because of their relatively poor scoring record (and emphasis on defence) in most of the years shown.

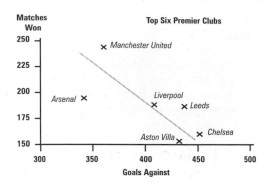

1.9 Gantt charts

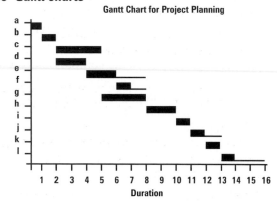

A Gantt chart gives a good visual picture of a project, so that the viewer can immediately see where critical areas lie.

1.10 Histograms and frequency polygons

1

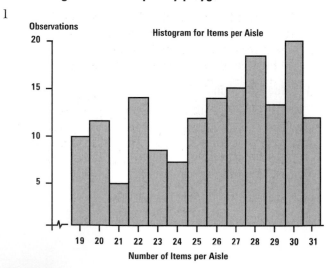

2

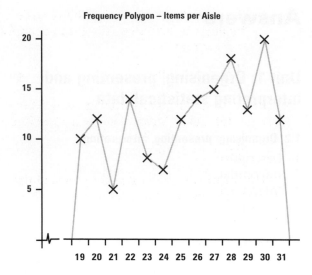

3 Either a histogram or a frequency polygon would be appropriate for showing this information. The histogram gives a clear comparison of different numbers of observations. The frequency polygon has a major advantage when comparisons need to be made, say between two supermarkets (numbers of items on aisles). On a histogram this would be confusing as some of the bars might coincide with one another.

Unit 2: Averages and Dispersion

2.1 The mean

1 a *The mean distance travelled by shoppers:*

40 shoppers travelled a total distance of 693 miles. The mean distance is therefore:

$$\frac{693}{40} = 17.3 \text{ miles}$$

b i) 16 shoppers with cars travelled a total distance of 535 miles. The mean distance is therefore:

$$\frac{535}{16} = 33.3 \text{ miles}$$

ii) 12 shoppers travelling by bus covered a total distance of 126 miles. The mean distance is therefore:

$$\frac{126}{12} = 10.5 \text{ miles}$$

c i) The total distance travelled by 18 male shoppers is 388 miles. The mean distance is therefore:

$$\frac{388}{18} = 21.5 \text{ miles}$$

ii) The total distance travelled by 22 women is 305 miles. Mean distance is therefore:

$$\frac{305}{22} = 13.8 \text{ miles}$$

c The mean is easy to calculate provided that there are not too many figures.

2.2 The median

1 A frequency table of customer responses shows that:

> 23 customers ranked the café excellent
> 18 customers ranked the café very good
> 4 customers ranked the café good
> 2 customers ranked the café average
> 3 customers ranked the café poor

As there were 50 responses, the median value is the 25th. The median value is therefore (ii) – very good. There are 23 (i)s and 18 (ii)s, therefore the median is just below (i), i.e. (ii).

2 The median represents the middle value and therefore is representative of an average value in the middle. It avoids the distortion caused by extreme values at either end of the range. It is also easy to calculate provided there are not too many values.

3 The median wage gives a good view of the typical wage of earners. By contrast, the mean can be misleading, because of the problem of extreme values influencing the rest.

2.3 The mode

1 The Sheffield United fan claimed that Sheffield Wednesday's average was zero because they had scored 0 in 6 matches. The next most popular score was 1, which they had scored 4 times.

2 The mean could have been calculated by dividing 16 by 14 = 1.14. The median could have been selected by lining up all 14 numbers (0 0 0 0 0 0 1 1 1 1 2 2 3 5) and choosing the middle number of the data set – 1.

3 It depends on the purpose of the average. In this case the mode is not at all helpful. The mean gives a good idea of the number of goals scored in a typical 90-minute period.

2.4 The range

1 The range is the difference between the highest and lowest value.

2 The range shows the extremes of the dispersion. However, it does not give a picture of where the bulk of the values lie within the range. For example 90% of the values could lie at the top of the range.

3 Extreme values tend to distort averages. For example, the fact that a few people earn millions of pounds in prize money from the National Lottery tends to distort the average figure that winners receive.

2.5 The quartile and interquartile range

1 a The upper quartile is the point separating the top two quarters of the range.

b The lower quartile is the point separating the bottom two quarters of the range.

c The interquartile range is the difference between the upper quartile and the lower quartile.

d There are ten deciles, each representing one tenth of the total distribution of values of a variable.

e There are 100 percentiles, each representing one hundredth of the total distribution of values of a variable.

2 The interquartile range gives a better picture of dispersion because it examines the majority of figures that lie close to the mean. It is therefore concerned with more typical values than are covered by the complete range.

2.6 Standard deviation

1 The range for Production Line 1 = 23–49. The range for Production Line 2 = 32–52.

2 The interquartile range is from 32–34 for Production Line 1, and from 33–34 for Production Line 2.

3 The standard deviation of Production Line 1 is 3.895 and for Production Line 2 is 2.987.

4 The standard deviation takes all the values into account, and is good when data is fairly symmetrical. It is a good way of checking for variations from the norm.

Unit 3: Probability

3.1 The probability of events occurring

1 a The probability is 2/10 = 0.2
  b The probability is 1/10 = 0.1
  c The probability is 3/10 = 0.3

2 a 0.1375
  b 0.3875
  c 0.3

3.2 The probability of combinations of events occurring

$$\frac{1}{5} \times \frac{1}{3} = \frac{1}{15}$$

3.3 Expected values

1 The **expected value** is the expected financial value of a business decision based on the probability of given financial profits or losses.

2
$$\begin{array}{ccc} & Success & Failure \\ \text{Expected Value} = 0.4 \times +£4m & + & 0.6 \times -£2m \end{array}$$
$$= £1.6m - £1.2m$$
$$= £0.4m$$

3.4 Decision Trees

1 In this case the decision is whether to spend £200m on the factory or to save the £200m. The expected value resulting from the three alternatives needs to be greater than 0. If we examine the three alternatives from the chance fork:

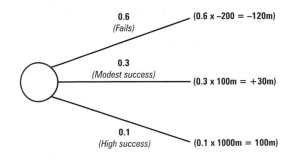

This gives an expected value of:

$$\frac{-£120m + £30m + £100m}{3} = +£3m$$

We can thus recommend that the business goes ahead with the decision.

2

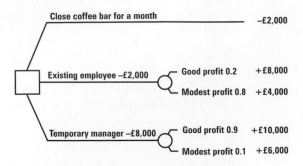

From the three alternatives:

• Closing the coffee bar will lead to an expected value of –£2,000.
• Using an existing employee will lead to an expected value of:

0.2 x £8,000 + 0.8 x £4,000/2 – £2,000

= £400

• Using a temporary manager will lead to an expected value of:

0.9 x £10,000 + 0.1 x £6,000/2 – £8,000

= –£3,200

The best alternative therefore is to use an existing employee and make a profit of £400.

3.5 Normal distribution

1 a The normal distribution is an 'idealised' curve in which the majority of values lie close to the mean, with fewer values further from the mean.
  b In a normal distribution, 68% of all observations lie within one standard deviation of the mean, and 95% of all observations lie within two standard deviations of the mean.

2 a 4 cents: 0.62%
    5 cents: 6.06%
    6 cents: 24.17%
    7 cents 38.30%
    8 cents 24.17%
    9 cents 6.06%
    10 cents 0.62%

  b The total price of 10,000 onions is 70,000 cents. This is calculated by multiplying the price of onions by the number of onions falling within each category, and summing the results.

c Mean price = 7p.

3 a

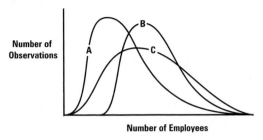

Number of Observations

Number of Employees

b Company C has a small number of employees earning much higher salaries than the rest.

c i Median salary = £20,000 – £21,999, probably about £21,000.

ii If the distribution is normal, you would expect 68% of all observations to lie within one standard deviation of the mean, i.e. between £19,000 and £23,000. This appears to be the case, assuming that salaries within the ranges £18,000–£19,999 and £22,000–£23,999 are dispersed in such a way that more lie close to the mean than further away from it.

iv Government statistics will provide an index of wage and salary increases in the UK, as well as indexes of price changes (the Retail Price Index). Employees will seek to match the normal standards for wage increases in order to keep pace with the cost of living.

3.6 Sampling

1 A representative sample is one which gives a true and fair picture of the total population. It can be achieved by using an extensive sample frame that is representative of the total population, often based on random selection across the frame.

2 The larger the size of the sample, and the more representative it is of the total population, the more likely it is to be close to a normal distribution.

Unit 4: Relationships Between Variables

4.1 Scatter graphs and lines of best fit

1

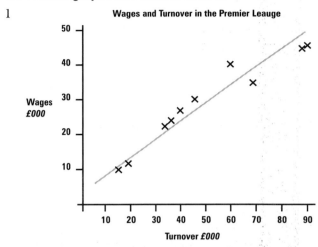

Wages and Turnover in the Premier Leaure

Wages £000

Turnover £000

2 The line of best fit gives a good general outline of the relationship between variables. Although some points may be dispersed away from the line, there is a tendency for values to lie close to it, which helps when predicting other values.

4.2 Correlation

1 Demand and Price: *Negative*
Profit and Turnover (provided costs are controlled): *Positive*
Price of Euro and Sales to USA: *Negative*
Hours worked to Output: *Positive*
Training to Faults Reported: *Negative*

2 A graph will show that significant drops in manufacturing output are accompanied by higher unemployment. As industry contracts, it sheds labour. More information about the causes of the decline in manufacturing may help to explain why unemployment is rising.

4.3 Regression

1,2

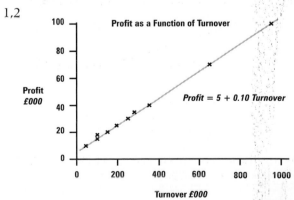

Profit as a Function of Turnover

Profit £000

Profit = 5 + 0.10 Turnover

Turnover £000

3 Profit = 5 + 0.10 turnover

4 • £35,000
  • £12,000
  • £65,000

4.4 Forecasting

1 Answers include:
  • The relationship between costs and prices
  • The relationship between profits and turnover
  • The relationship between wages and output.
2 If the firm can forecast the relationship between profits and turnover, it will have a better idea of how much it will need to sell in order to achieve targeted profit figures.
3 Forecasts are based on previous information. In the course of time, the relationship between variables alters and there are many random affects that cannot be predicted.
4 T = 2C. If there is 'always' this relationship, then the probability is 1. However, in the real world, there will be all sorts of factors which lead to unpredictability.
5 The chart shows a fall both in the growth of the market and in the value of sales from the late 1990s onwards. The growth figure is now negative, indicating that the market has gone beyond maturity into decline, particularly as more people pirate music off the Internet. It is possible that this decline will continue at least in the short term (i.e. the next year or so) and possibly also in the longer term. However, this forecast will need to be reviewed in the light of developments in the market.

Unit 5: Time Series

5.1 Underlying trends

1 One month's figures are not enough to show a trend, unless they are part of an ongoing time series which has been put together by working out the moving average. To identify a trend, it is necessary to develop a time series and to remove from it any seasonal and other variations.
2 Remember to centre the values: 2.6, 2.9, 3.3, 4.0, 4.7, 5.6, 6.5.
  The mean between 94 and 96 is 3.5; between 98 and 00 is 9.0, and between 94 and 00 is 6.5.

5.2 Seasonal variation

1 A 'season' is made up of a typical sequence of periods, e.g. four seasons in a year, or five working days in a week, etc.
2 In summer and winter the graph would need to indicate a rise of 10 above the trend, which is rising at 10 per month.

5.3 Seasonally adjusted figures

1 The figures do not account for seasonal variations. For example, each year unemployment figures may fall in April, May and June as more casual work becomes available on building sites. In July and August the jobless total is swollen by school leavers and students entering the job market.
2 Taking a moving average based on the 12 months of the year, or the four seasons, will smooth out the seasonal variations.

5.4 Forecasting future figures

Forecast Demand

May	50,000
June	51,000
July	52,000
August	53,000
September	54,000
October	55,000
November	56,000
December	59,000
January	61,000
February	62,000
March	63,000
April	64,000
May	65,000

5.5 Cyclical and random variables

1 The trade cycle is made up of regular periods of upturns and downturns in economic activity. Although these periods succeed each other, it is difficult to predict the length and extent of booms and slumps in activity.

By contrast, seasonal trends appear at regular intervals, and it is easy to predict the seasonal impact using an additive or multiplicative approach.
2 Random factors are normally discounted because they are not part of a trend. A trend has some sort of regular pattern to it.

Unit 6: Indices

6.1 Simple index numbers

1 A typical index is set out as:

$$\frac{Vn}{Vo} \times 100$$

where Vn represents the value in a particular year (n) and Vo represents the base-year value (0).

2 a

	2000	2001	2002	2003	2004
Bread	100	106.25	112.5	118.75	125
Cheese	100	120	130	140	160
Eggs	100	108.33	116.66	125	133.33
Wine	100	105	110	115	120
Cigarettes	100	102.5	110	112.5	125

b *Composite index:*

2000	2001	2002	2003	2004
100	108.4	115.8	122.2	132.0

6.2 Complex index numbers

1 Weighting makes an index more representative because it reflects the relative importance of different items covered by the index.
2 It may be helpful to weight items in a way that is more representative of current patterns. For example, in the case of the RPI, it would be pointless to assign weights based on 20 years ago – some items that were significant then may not be at all significant today.
3 129.
4 This would necessitate a change in weighting. The index is currently weighted in terms of the first (base) year.
5 You could use the second year as the base for the weighting.
6 a 124.2.
  b 119.4

6.3 Important published indexes

1 The FTSE All Share Index is the most representative of all share price indices, reflecting the near-800 companies appearing in the *Financial Times* (FT) share price list.
2 The FTSE 100 (or 'Footsie') started with a base of 1,000 at the end of December 1983 and reflects price movements of the most significant 100 shares listed on the London Stock Exchange.
3 The FTSE All Share will fluctuate more because a number of its shares are more volatile. The top 100 companies have more stability and the public has more confidence in them as a result.
4 a Launched in 1992, the FTSE 250 Index comprises the range of companies immediately below the size that qualifies for the FTSE 100.
  b The FTSE 100 did better because it lists 'blue chip' companies which lie at the core of the economy and thus are less prone to speculation.
  c Points reflect movements in the index rather than in actual share prices: they give an overall index of changes that are taking place.

Glossary

Bar chart Type of graph used to display and compare number, frequency or other measures (e.g. mean) for different, discrete categories of data.

Bi-modal Where there are two values which are the most 'popular' or common in a series.

Business cycle Pattern of fluctuations in business activity, with alternating periods of increased and reduced activity.

Census Periodic survey of population and its characteristics. In the UK a major census is carried out every ten years.

Correlations Mutual relation between two or more items.

Correlation coefficient Numerical measure of the strength of correlation.

Deviation Amount by which a single measurement differs from the mean.

Dispersion Extent to which values in a series differ from the mean.

Distribution Pattern according to which a characteristic is spread over members of a class.

Frequency Measure of occurrence of particular events or observations.

Histogram Form of bar chart in which data represent continuous rather than discrete categories. In a histogram, there are no gaps between the columns representing the different categories, and the area of the bar is proportional to the size of the category.

Horizontal bar chart Chart with horizontal bars, the length of which indicates size of category.

Index number Number showing variation of prices, wages or some other measured quantity, compared to a chosen base period.

Inference Conclusion based on certain premises or information.

Inter-quartile range Measure indicating extent to which the central 50% of values within a data set are dispersed.

Line of best fit Line drawn on a graph to indicate central tendency.

Mean Arithmetical average calculated by adding together all the values in a series and dividing the total by the number of observations.

Median Middle value of a series of values arranged in order of size.

Mode Value that occurs most frequently in a given set of data.

Normal distribution Function representing the distribution of variables as a symmetrical bell-shaped graph.

Pie chart Circular graph that shows the relative contribution that different categories make to an overall total.

Probability Likelihood of an event occurring, measured by the ratio of the cases to the total number of cases possible.

Range Difference between the highest and the lowest value in a data set.

Scatter plots Dots plotted to show the relationship between pairs of quantitative measurements made from the same object or individual.

Segment Sector or part of a total population.

Standard deviation Quantity calculated to indicate the extent of deviation for a group as a whole; a summary of the amount by which every value within a data set varies from the mean.

Trend General direction or tendency

Variable Quantity able to assume different numerical values

Vertical bar chart Chart with vertical bars, the height of the bar indicating the size of the category.

X-axis Horizontal line defining the base of a plot area. Depending on the type of graph, different locations on the x-axis represent either different categories (such as years) or different positions along a numerical scale (such as income).

Y-axis Vertical line usually defining the left side of the plot. However, if more than one variable is being plotted on the graph, the vertical lines on both the left and right side of the plot area may be used as y-axes.

Index

Note: Entries in italic are included in the Glossary on page 101

Addition law of probability 42–45
Additive approach 74
Averages 24–33

Bar charts 9
Base weighting 86
Base year 84–90
Bi-modal distribution 27
Boston Consultancy Group 14
Business cycle 81

Central tendency 59
Chance forks 44–48
Class intervals 18, 19
Correlations 60–65
Correlation analysis 60
Correlation coefficient 61
Cumulative frequency 19

Deciles 31
Decision forks 44–48
Decision trees 44–48
Dependent variable 62
Descriptive statistic 3
Deviation 32–33
Dispersion 28–31
Distribution curves 49–54
Domesday Book 2
Dow Jones 89

Expected values 43–48
Experience 14

Family Expenditure Survey 53, 90
Forecast 64–65
Frequency distribution 4–6
Frequency polygon 19
Frequency tables 6, 17, 25
FTSE indices 89

Gantt, Henry 15–16
Gantt charts 15–16
Graunt, John 2–3

Histograms 17–18

Independent variable 62
Inferential statistics 3
Intercept 63
Inter-quartile range 30–31

Inventory 4
Least squares 63
Linear array 5
Line graphs 12–13
Line of best fit 59–64
Linked probability 42–43

Mean 24–33
Median 24–26
Mode 27
Moving average 73–79
Multi-modal distribution 27
Multiplication law of probability 41–43
Multiplicative approach 74–76

National Lottery 54
Negative correlation 60
Negative skew 49
Negative variation 75
Nikkei 89
Normal distribution 50–54

Pearson correlation coefficient 61
Percentage changes 85
Percentiles 31
Pictogram 8
Pie charts 10
Plotting trends 76
Point of inflection 52
Population census 53
Positive correlation 60
Positive skew 49
Positive variation 75
Prediction 80
Price relatives 85–88
Probability 37–56
Probability distribution 38–56
Pruning back 48

Quartiles 31

Random selection 54
Random variation 72–81
Range 5, 28–33
Regression 65
Regression line 63
Relative frequency 6, 39–40
RPI 90

Salaries (average national) 7
Sampling 4, 53–54
Sampling error 53
Sampling frame 54
Scatter graphs 13, 58–65
Seasonal adjustment 79
Seasonal variation 70–81
Segments 10
Semi-averages 72
Skew 49
Smoothing 78–81
Split plots 65–66
Standard deviation 32–33, 51–54
Statistic (defined) 2

Tables 5–11
Terms of trade 84
Time series 11–13
Trends 11–13, 70–80

Underlying trends 71–80

Variance 32–33
Variation 28–33

Weighting 86–90